TESTING STATIC RANDOM ACCESS MEMORIES

FRONTIERS IN ELECTRONIC TESTING

Consulting Editor
Vishwani D. Agrawal

Books in the series:

TESTING STATIC RANDOM ACCESS MEMORIES

DEFECTS, FAULT MODELS AND TEST PATTERNS

by

SAID HAMDIOUI
Delft University of Technology, The Netherlands

KLUWER ACADEMIC PUBLISHERS

BOSTON / DORDRECHT / LONDON

A C.I.P. Catalogue record for this book is available from the Library of Congress.

ISBN 978-1-4419-5430-5

Published by Kluwer Academic Publishers,
P.O. Box 17, 3300 AA Dordrecht, The Netherlands.

Sold and distributed in North, Central and South America
by Kluwer Academic Publishers,
101 Philip Drive, Norwell, MA 02061, U.S.A.

In all other countries, sold and distributed
by Kluwer Academic Publishers,
P.O. Box 322, 3300 AH Dordrecht, The Netherlands.

Printed on acid-free paper

To my parents

Contents

Preface

Semiconductor memories, which are an integral part of any modern ULSI chips (e.g. microprocessor, ASICs, etc.), are widely considered to be one of the most sensitive and critical component of current digital systems. Not only because the memory share of the chip area increases and is expected to be about 94% in 2014, but also because of technology shrinking, which makes memories more sensitive for defects. The challenge of testing and diagnosing semiconductor memories grows, especially as the semiconductor technologies and the level of integration continue to improve. Precise fault modeling and efficient test design, in order to keep test cost and time within economically acceptable limits, is therefore essential. The quality of the used tests, in terms of their fault coverage and test length, is strongly dependent on the used fault models.

This book deals with the study of fault modeling, testing and test strategies for semiconductor *Static Random Access Memories (SRAMs)*; including single-port multi-port SRAMs. The major goal of this book is to contribute to the technical acknowledge needed by those involved in memory testing, engineers and researchers. The book does this by presenting a systematic development of fault models, optimal test algorithms and test strategies. A complete framework for memory faults for single-port and dual-port SRAMs is presented, using the concept of *fault primitive*, which is a compact mathematical description that exactly describe a faulty memory behavior. Special attention has been given to defect injection and SPICE simulation in order to verify the reality of the framework for single-port, two-port and three-port memories; all possible defects in the memory cell array are considered and modeled as a resistor with a variable value. A systematic way to develop test patterns for the established realistic models is presented, together with the tests targeting the considered faults as well as test strategies. Some industrial test results are reported in order to investigate the quality of the introduced test techniques as compared with the traditional ones. The impact of the port restrictions (e.g., read-only ports, write-only ports) on the fault models, tests and test strategies is also covered.

Organization

This book is organized into parts; each part consists of one ore more chapters. Appendices have been included to contain details which contribute to the completeness of the material, but not essential to the overall understanding.

Part I. Introductory.

The purpose of this part is to provide a general introduction, background and a preparation for simulation. It consists of Chapter 1 through Chapter 4.

Chapter 1 highlights briefly the technological developments in Ultra Large Scale Integrated (ULSI) circuits, from which the increased importance of testing follows. It also introduces the existing memory types and competing technologies, and describes semiconductor memories. In addition, it establishes the state-of-the art of SRAM testing, and represents the research topics.

Chapter 2 discusses the structure of single-port (SP) and multi-port (MP) SRAMs, by using the modeling approach in a top down fashion. Each description level is concerned with specific aspects of the SRAM, beginning with its behavior down to the physical buildup of its components. First, the chapter provides the external SRAM behavior treated as a black box, with only the input and output lines visible. Then, it introduces the functional SRAM model, along with a description of some real SRAM devices. Thereafter, it describes the electrical circuits that make up the functional blocks. Finally, it treats the SRAM cell layouts.

Chapter 3 defines the complete space of the memory faults for SP and two-port (2P) memories. It first defines the concept of fault primitives and functional fault models. Then it classifies the faults in SRAMs and shows the scope of the faults targeted in this dissertation. Thereafter, it presents a complete set of SP faults and 2P faults.

Chapter 4 discusses the preparation needed to begin with the simulation in the next stages of the dissertation. The simulation is used to investigate the validity of the set of faults introduced in Chapter 3. Chapter 4 starts with defining and locating the defects to be simulated in a MP memory cell array. Then, it introduces an appropriate simulation model, which accurately describes the behavior of the memory, and while only requiring a reasonable simulation time. Thereafter, it presents the simulation methodology to be followed during the simulation. Finally, it gives some simulation results for the defect free case.

Part II. Testing two-port SRAMs.

This part concerns with developing realistic fault models, tests and test strategies for SP and 2P memories. In addition, it covers the impact of port restriction (e.g., read-only port, write-only port) on the fault models as well as on the tests for 2P memories. It consists of Chapter 5 through Chapter 7.

Chapter 5 concerns with the circuit simulation for 2P memories. It first introduces the list of defects to be simulated. Then, it presents electrical simulation results, which thereafter are translated into realistic FFMs. Furthermore, the chapter gives the probability of occurrence of the introduced FFMs based on two approaches; first by assuming that all modeled defects are equally likely to occur, and thereafter by performing inductive fault analysis on two different layouts representing the same electrical circuit.

Chapter 6 establishes march tests for SP and 2P memories. It starts first by stating the notation for march tests. Then, it introduces a new single-port test detecting realistic single-port faults derived (based on circuit simulation) in Chapter 5, and a test detecting all possible single-port faults described in Chapter 3. Thereafter, it develops conditions for detecting unique two-port faults (2PFs); such conditions are used to derive tests. The chapter also gives the test strategy, and presents industrial test results.

Chapter 7 discusses the impact of port restrictions on the FFMs introduced in Chapter 5, and on the tests and the test strategy presented in Chapter 6. It first classifies the 2P memories, depending on the type of ports they consist of (i.e., read-only, write-only or read-write port). Thereafter, it establishes the impact of such restrictions on the FFMs, on the tests, as well as on the test strategy.

Part III. Testing p-port SRAMs.

This part concerns with developing realistic fault models, tests and test strategy for three-port (3P) memories; and extending the results to any MP memory with p ports. It also covers the impact of port restriction on the introduced FFMs as well as on the tests. It consists of Chapter 8 through Chapter 10.

Chapter 8 concerns with the circuit simulation for 3P memories. It first introduces the list of defects to-be simulated. Then, it presents electrical simulation results. Such results are then translated into realistic FFMs. The occurrence probability of each FFM is also calculated by assuming that all modeled defects are equally likely to occur. Furthermore, the chapter extends the results to any multi-port memory with p ports.

Chapter 9 establishes march tests for any p-port memory, together with an optimal test strategy. It first starts with introducing conditions to detect the p-port faults $(p > 2)$; such conditions are used to derive tests. The chapter also presents an optimal test test strategy.

Chapter 10 deals with the impact of port restrictions on the FFMs, the tests and the test strategy for p-port memories $(p > 2)$. It first classifies the p-port memories into p classes, depending on the ports the memory consists of. Thereafter, FFMs for each class together with their tests and test strategy is discussed.

Chapter 11 presents the further research challenges and trends in testing (embedded) memories.

Appendix A gives all simulation results, for all defects simulated in 2P SRAMs.
Appendix B lists all simulation results, for all defects simulated in 3P SRAMs.

Acknowledgements

The material presented in this book contains the major results of five year research done at the Computer Engineering (CE) Laboratory, Faculty of Electrical Engineering, Mathematics and Computer Science, Delft University of Technology, The Netherlands; and at the Architecture group of Intel Corporation at Santa Clara as well as at Folsom, California, USA. Numerous people have contributed in several ways to establish the results presented here. It would go to far to mention all of them; however, I would like to take the opportunity to thank some of them.

First of all, I like to thank Prof. Ad van de Goor who took the first effort in introducing me the the world of memory testing. I'm grateful to him for his guidance and help for writing this book, and for his feedback and many comments and ideas. Furthermore, I'd like to thank Prof. Stamatis Vassiliadis, for his continue moral support and voluntary advise not only in research area but also in many other fields of life. Thanks to many colleagues reviewed the book or individual chapters, especially: dr.ir. J. Geuzerbroek, ir. Z. Al-ars and ir.G.N. Gaydadjiev. Thanks to all CE members who jointly created a very friendly and flexible work environment for a successful research.

Moreover, I would like to thank the involved people from Intel Corporation at Santa Clara and Folsom, California, USA; especially, Mike Rodgers, Greg Tollefson, Mike Trip, Bob Roeder, John Reyes and others. Thanks for providing the designs, the tools, as well as for your feedback, comments, and suggestions. Special thanks to David Eastwick for his continued support and comments.

But most important, I would like to thank my parents and all my family and friends, who have always stood by me, for their moral encouragement and support. I could never have done this without your support; for all of you, *tkathig kanniou attas*.

Mostly, thanks are due to God.

Said Hamdioui   Rotterdam, October, 2003
⊙·ΠΣΛ λ· ⊏ ΛΣ⊔Σ   The Netherlands

Symbols and notations

0 denotes a zero value

1 denotes a one value

x denotes a 1 or a 0 value; $x \in \{0, 1\}$

d denotes a don't care value

? denotes a undefined, unknown or random value

$0r0$ denotes a read 0 operation from a cell containing 0

$1r1$ denotes a read 1 operation from a cell containing 1

$0w1$ denotes an up transition write operation

$1w0$ denotes a down transition write operation

$0w0$ denotes a non-transition write 0 operation

$1w0$ denotes a non-transition write 1 operation

xOy denotes any possible operation O; $xOy \in \{0r0, 1r1, 0w1, 1w0, 0w0, 1w1\}$

"$xO_1y : zO_2t$" denotes two simultaneous operations xO_1y and zO_2t

"$xO_1y : zO_2t : \ldots : mO_pn$" denotes p simultaneous operations xO_1y till mO_pn

$< S/F/R >$ denotes a FP where S is a sensitizing operation sequence, F is the fault effect, and R is the read value in case S applied to the victim cell is a read operation

$< fault_1 > \& < fault_2 > \&\ldots\& < fault_p >$ denotes a p-port FFM consisting of p weak faults; e.g., wRDF&wRDF consists of two weak RDF's

⇑ denotes an up addressing order

⇓ denotes a down addressing order

⇕ denotes an irrelevant addressing order

1PF: Single- (i.e., one-)port faults

1PF1: 1PF involving a single cell

1PF2: 1PFs involving two cells

$1PF2_s$: 1PF2 where the state of the aggressor cell sensitize a fault in the victim cell

$1PF2_a$: 1PF2 where an operation applied to the aggressor cell sensitize a fault in the victim cell

$1PF2_v$: 1PF2 where an operation applied to the victim cell sensitize a fault in that victim cell, with the aggressor cell is in a certain state

2P: two port

2PF: two-port fault

2PF1: 2PF involving a single-cell

2PF2: 2PF involving two cells

$2PF2_a$: 2PF2 where two simultaneous operations applied to the aggressor cell sensitize a fault in the victim cell

$2PF2_v$: 2PF2 where two simultaneous operations applied to the victim cell sensitize a fault in that victim cell, with the aggressor cell is in a certain state

$2PF2_{av}$: 2PF2 where two simultaneous operations, one applied to the aggressor cell and one applied to the victim cell, sensitize a fault in that victim cell

2PF3: 2PF involving three cells

3P: three port

3PF: three-port fault

3PF1: 3PF involving a single cell

3PF2: 3PF involving two cells

$3PF2_a$: 3PF2 where three simultaneous operations applied to the aggressor cell sensitize a fault in the victim cell

$3PF2_v$: 3PF2 where three simultaneous operations applied to the victim cell sensitize a fault in that victim cell, with the aggressor cell is in a certain state

pP: p-port

pPF: p-port fault

pPF1: pPF involving a single cell

pPF2: pPF involving two cells

pPF2$_a$: pPF2 where p simultaneous operations applied to the aggressor cell sensitize a fault in the victim cell

pPF2$_v$: pPF2 where p simultaneous operations applied to the victim cell sensitize a fault in that victim cell, with the aggressor cell is in a certain state

a-cell: aggressor cell

CF: coupling fault

CFdrd: Deceptive Read Destructive CF

CFds: Disturb CF

CFid: Idempotent CF

CFin: Inversion CF

CFir: Incorrect Read CF

CFrd: Read Destructive CF

CFrr: Random Read CF

CFrrd: Random Read Destructive CF

CFst: State CF

CFtr: Transition CF

CFud: Undefined Disturb CF

CFur: Undefined Read CF

CFus: Undefined State CF

CFwd: Write Destructive CF

CFuw: Undefined Write CF

DRF: Data Retention Fault

DRDF: Deceptive Read Destructive Fault

FFM : functional fault model

FP: fault primitive

IRF: Incorrect Read Fault

MP: multi-port

NAF: No Access Fault

Pro: read-only port

Prw: read-write port

Pwo: write-only port

RDF: Read Destructive Fault

RRDF: Random Read Destructive Fault

RRF: Random Read Fault

SAF: Stuck-At Fault

SOS: sensitizing operation sequence

SP: single-port

SF: State Fault

TF: Transient Fault

v-cell: victim cell

USF: Undefined State Fault

URF: Undefined Read Fault

UWF: Undefined Write Fault

$wFFM_i$: denotes a weak FFM_i; e.g., wRDF denotes a weak RDF

WDF: Write Destructive Fault

Part I

Introductory

Chapter 1. Introduction

Chapter 2. Semiconductor memory architecture

Chapter 3. Space of memory faults

Chapter 4. Preparation for circuit simulation

This part is to provide a background and a preparation for simulation. It consists of four chapters. Chapter 1 gives some basics about testing, and discusses the state-of the art in SRAMs testing. Chapter 2 describes SRAMs architecture in a top down fashion. Each description level is concerned with specific aspects of the SRAM, beginning with its behavior down to the physical buildup of its components. Chapter 3 defines the whole space of memory faults based on the 'fault primitive' concept. Chapter 4 discusses the preparation needed to begin with simulation in the next stages of the book.

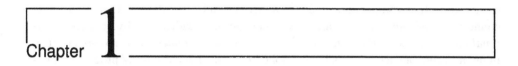

Chapter **1**

Introduction

1.1 Test philosophy
1.2 Memory Technology
1.3 Modeling and testing faults in SRAMs

Fast and efficient testing is an important step in any manufacturing process. With the recent advances in semiconductor technology, the design and use of memories for realizing complex systems-on-a-chip has been widespread. The cost of testing such memories increases rapidly with every generation. Precise and realistic fault modeling, and efficient test design, in order to keep test cost and time within economically acceptable limits, are therefore essential.

This introductory chapter starts out showing the importance of testing and some of its basics. Then, it introduces the existing memory types and competing technologies, and describes semiconductor memories. Finally, it discusses the state-of the art in fault modeling and test design for SRAMs, and the remain research topics.

1.1   Test philosophy

Ultra large scale integrated (ULSI) circuits are an integral part of any modern electronic system. Such circuits contain from millions to over one billion transistors, diodes and other components such as capacitors and resistors, together with interconnections, within a very small area. They can be divided into two classes: *combinational circuits* (without memory) and *sequential circuits* (with memory). The manufacturing of such circuits is a complicated and time-consuming process and defects in them are inevitable. Such defects may be due to several deficiencies in the original silicon and in the manufacturing process. Examples of the former are impurities and dislocations, and examples of the latter are temperature fluctuations during wafer processing, open interconnections, open circuits, short circuits, and extra or missing transistors. The complexity of ULSI technology has reached the point where chips already contain over 1 billion transistors on a single chip, with on-chip clock frequencies of over 1GHz [77]. These trends have a profound effect on the cost and difficulty of chip testing. The larger the ULSI circuit in terms of area and complexity, the greater the chances of it having a defect. From the point of view of economics, it has been shown that the cost of detecting a faulty component is lowest before the component is packaged and becomes part of a ULSI system. Therefore, testing is a very important aspect of any ULSI manufacturing process.

As mentioned previously, ULSI circuits can be classified into combinational circuits and sequential circuits. Therefore, a distinction can be made between testing each one of the two classes. Testing combinational circuits is much easier, since for each fault, one or two test vectors have to be applied; while the detection of faults in sequential circuits may require first to bring the circuit into a state in which the fault may be sensitized and observed. For example, in a sequential circuit with a state vector of n bits, this might require 2^n input signals (or clock cycles) before the faults will be detected.

There are two aspects of ULSI testing: *fault detection* and *fault diagnosis*. The testing process involves the application of test patterns to the circuit and comparing the response of the circuit with a precomputed expected response. If a product is designed, fabricated, and tested, and fails the test, then there must be a certain cause for the failure [13]. Either (a) the test was wrong, (b) the fabrication process was faulty, (c) the design was incorrect, or (d) the specification had a problem; anything can go wrong. The role of fault detection is to detect whether something went wrong; while the role of fault diagnosis is to determine exactly what went wrong and where the process needs to be altered. Therefore, the correctness and the effectiveness of testing is most important for perfect products.

ULSI testing can be classified into four types, depending on the purpose it accomplishes [86]: *characterization, production, burn-in*, and *incoming inspection.*

Characterization: also known as *design debug* or *verification testing.* This form of testing is performed on any new design before it is sent to production. It has to verify that the design is correct and meets the specifications; it also has to determine the exact limits of the device operating values. *Functional tests* are run during that phase, and comprehensive AC and DC parametric measurements are made. Tests are generally applied for the worst case, because they are easier to evaluate than average cases and devices passing these tests will work for any other conditions. Probing of the internal nodes of the chip may also be required during the design debug.

Production: every fabricated chip is subjected to production tests, which must enforce the relevant quality requirements by determining whether the device meets the specifications; they are less comprehensive than characterization tests. The test may not cover all possible functions; however, they must have a high fault coverage for the modeled faults. Fault diagnosis is not attempted and only a pass/fail decision is made.

Burn-in: this ensures the reliability of devices, which pass production tests, by testing either continuously or periodically over a long period of time at elevated voltage and temperature [38]. Burn-in causes bad devices to actually fail. Two types of failures are isolated by burn-in: *infant mortality and freak failures.* Infant mortalities are screened out by a short term burn-in, typically 10-30 hours; they are often caused by a combination of sensitive design and process variation. Freak failures are devices having the same failure mechanisms as the reliable devices, but requiring long burn-in time, typically 100-1000 hours. During burn-in, production tests are applied at high temperatures and with an over-voltage power supply.

Incoming inspection: incoming inspection is performed on purchased devices, before integrating them into a system. The most important purpose of incoming inspection tests is to avoid placing a defective device in a system, where the cost of diagnosis may far exceed the cost of incoming inspection.

Another classification for ULSI testing can be made depending on the type of targeted faults; either *parametric* or *functional.*

Parametric testing is necessary to verify whether the chip meets DC and AC specifications. The DC parametric tests include maximum current, leakage, output drive current, thresholds levels, etc.; while the AC parametric tests include propagation delay, setup and hold times, functional speed, access time, and various rise and fall times.

On the other hand, functional testing determines whether the internal logic function of the chip behaves as intended; it checks for a proper operation of the design. Such test has to guarantee a very high fault coverage of the modeled faults. Fault modeling of physical faults is a very important aspect in functional testing, since it turns the problem of test generation into a technology-independent problem. In addition, tests designed for modeled faults may be useful for detecting physical faults whose effect on circuit behavior is not well understood and/or too complex to be analyzed otherwise. *Fault modeling* and *test generation* for *multi-port memories* are the main topic of this thesis.

1.2   Memory technology

In this section, first a general overview of existing memory types and the competing technologies is given. Thereafter, the semiconductor memories will be described shortly.

1.2.1   Memory classification

A memory device, as its name may suggest, is a means to store and retrieve information. A bit of information (logic 1 or logic 0) is stored in a *memory element*, or *memory cell*. The most straight forward approach to produce a device that stores more than one data bit is to arrange a number of memory elements in an array, and give each element a name (also called an address).

In the past five decades, several concepts were proposed to implement memory devices, with different degrees of success. These different implementations may be classified in many ways, based on their performance, principle of operation, architecture, technology, etc. Figure 1.1 shows a classification of the most widely used and the most promising memory devices, based on their principle of operation. As seen in the figure, memories are written or read either by electrical, magnetic or optical signals. Such a classification allows for the comparison of the different information storing schemes, without the need to discuss the architectural aspects of each specific design. Generally speaking, magnetic storage is the oldest, while the optical is the most recent scheme.

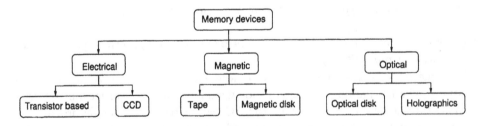

Figure 1.1. Classification of the most popular memory types

Optical devices

Optical devices are divided into holographic and optical disk memories. The idea of a holographic memory has not yet been harnessed on a mass production scale, but it promises to provide high storage and performance capabilities for future mainframes. The holographic memory stores data as holographic images in crystals by locally changing the refractive index of the crystal using a laser beam. By changing the laser angle of incidence, multiple data bits can be stored in the same region. This makes the high storage capacity possible.

Optical disks, on the other hand, are already being used massively, mainly as a read-only memory (ROM) for archival purposes. Because they have a high storage capacity, are easy to transport, and are relatively inexpensive, optical disks became the best choice to transport programs and audio files stored as dozens of megabits of encoded information from producers to consumers. Optical disk devises store data as series of dots and spaces carved on a plastic surface. Data bits are read by detecting the reflection of a laser beam incident on the surface of the rotating disk. One big disadvantage of optical disk storage is that it is relatively slow. Moreover, because it uses mechanical parts for rotation, it is less reliable than other non-mechanical forms of memory devices, even though their high capacity and low price ensure that they stay in competition for the time being.

Magnetic devices

The second type of storage devices is based on magnetic recording and retrieval of data. This type of data storage is one of the oldest in the computer memory industry. Magnetic storage is so popular that it held, and still holds, the highest memory market share amounting to about 60% of its total revenue. A large number of memory devices have been developed based on the principle of magnetic storage, but only the tape and the magnetic disk survived the market competition till today. Magnetic tape storage is a memory device used to store vast quantities of information. The technology used to implement it has not changed much since the early days of its development. For writing data, the magnetic tape recorder uses an inductive head that transforms incoming electrical signals into magnetic fields which are in turn registered on a moving tape covered with a magnetic material. The reading of data is the reverse operation where the head detects the magnetic field characteristics stored on the moving tape and transforms them back into electrical signals. Although magnetic tapes store large amounts of data at a very low cost, their performance is very low due to the requirement that the tape remains in direct contact with the inductive head. Therefore, if the tape were to run continuously at a high speed, the tape would soon wear out. The solution to this problem has lead to the introduction of the magnetic disk memory.

The key development that led to the magnetic disk technology was the invention of a low mass slider carrying an inductive head floating at a precise spacing between the head and the magnetic medium [9]. The data recording and retrieval mechanism is basically the same as that of the magnetic tape. Continued innovative improvements to the magnetic disk have made it the best secondary memory and memory backup storage medium for computers; e.g., hard disks and floppy disks. Magnetic disks managed to achieve a good balance between speed, capacity and price. They are thought to keep dominating the secondary memory and the backup market well into the next decade.

Electrical devices

The third type of storage devices is the electrical memory devices, better known as semiconductor memories; they are the most celebrated type of memories. The least known in the computer industry are the charge-coupled devices (CCDs). A CCD memory element operates by controlling the voltage level on both sides of the element, thus controlling the flow of electrical charge to and from the element. The logic level within an element depends on the amount of electrical charge contained in it. To access data in a given element, the electrical charge in this element needs to be transported from one cell to the other until it reaches a port where it is read; this is called *serial accessing*. Serial accessing is the main drawback of CCDs because it is slower than *random accessing*, which means the ability to access any cell in a memory device at random. This disadvantage forced CCDs out of the computer memory market, but they are still used in applications where high performance is not required, such as imaging in video cameras.

The other type of electrical memories is transistor based (i.e., *semiconductor memories*), which is the one most closely related to the computer industry. We dedicate the following section to the treatment of this important type of memories.

1.2.2   Semiconductor memories

Semiconductor memory devices have the highest performance and the highest versatility among all types of memories. Such memories, built from transistors, may be classified according to their architecture, into *random access memories (RAMs)* and *read-only memories (ROMs)*. These architectures may in turn be divided into designs with more specific characteristics.

The name RAM stands for a memory device in which any cell may be accessed at random to perform a read or a write operation. Depending on the internal architecture and the actual memory cell structure, RAMs may be further divided into *dynamic RAMs (DRAMs)* and *static RAMs (SRAMs)*.

On the other hand, ROMs are preprogrammed memory devices that are set to produce the same output at all times; unless altered to change the data placed within them. Depending on the erasure method, ROMs have two variants: *erasable, programmable ROMs (EPROMs)* which are erasable with ultraviolet (UV) light; and *electronically, erasable ROMs (EEPROMs)* which are erasable electronically. Table 1.1, adapted from [67], gives the most interesting properties for the different transistor based memory architectures; while Table 1.2 documents the memory market as away from EEPROM and EPROM and toward DRAMs and SRAMs [13].

Table 1.1. Summary of the characteristics of the different memory architectures

Criterion	DRAM	SRAM	ROM	EPROM	EEPROM
Relative cell size	1.5	4-6	1	1.5	3-4
Volatility	yes	yes	no	no	no
Data retention	64ms	∞	∞	10 years	10 years
In-system re-programmability	yes	yes	no	no	yes

Table 1.2. Semiconductor memory shares %

Year	1988	1990	1994
DRAMs	56	54	58
SRAMs	17	22	21
ROMs	8	8	6
EPROMs	17	14	12
EEPROMs*	2	2	3

*Includes flash EEPROM

RAMs

RAMs have been divided into DRAMs and SRAMs. DRAMs have their information stored as a charge on a single capacitor. They have the highest possible density, but a slow access time (typically $20ns$). Inherently, DRAM cells suffer from charge leakage, a phenomenon known as *leakage currents*, that cause a cell to lose its charge gradually. In order to maintain the state of a cell, it is necessary for DRAMs to rewrite, or *refresh*, the already stored data bits from time to time (typically every 64 ms) before the state cannot be recovered with certainty. The word 'dynamic' in the name of the device refers to the fact that the data stored in the DRAM cell has to be refreshed after a given period of time.

SRAM cells are constructed out of bistable electrical circuits, which means circuits that have two different stable states. Each state is used to represent a given logical level. Once a cell is forced into one of the two states, it will stay in it as long as the memory is kept connected to the power supply; the name 'static' refers to this property. Note that SRAMs do not require to be refreshed; however, they require more silicon area per bit than DRAMs. SRAMs have the fastest possible speed (typically $2ns$).

Both DRAMs and SRAMs are called *volatile devices* because they can only keep their data content if they stay connected to the power supply. A closer look at the two RAM structures reveals that, because SRAM cells are always in a given stable state, they consume less power than DRAM cells. Moreover, since DRAMs depend on capacitive elements and SRAMs do not, SRAMs have higher performance than DRAMs and are therefore used as the first level memory directly supporting processor units. The main advantage DRAMs have over SRAMs is in their density aspect. This means that DRAMs have a lower price per bit than SRAMs. The architecture of SRAMs will be discussed in more detail in Chapter 2.

ROMs

The other architectural variation of the transistor based memory is the ROM. These are preprogrammed memory devices that permanently store information needed by the processor during normal operation, such as look-up tables and code conversions. ROMs are non-volatile and, therefore, keep their stored data even if the power is turned off.

One variant of the ROM is the programmable ROM or PROM, which is delivered before programming to the consumer who freely uses it to store the information that suits his own application. The stored information cannot be erased, a property that puts a limitation on the reusability of the PROM if the application for which the memory is used changes.

To tackle this problem, the erasable PROM (EPROM) was introduced. Once programmed, the EPROM acts like a normal ROM. If needed, the user is capable of erasing the contents of the EPROM and can reprogram it. The disadvantage of the EPROM approach is that it cannot be reprogrammed while residing in the system, but must be removed from the system and erased with an ultraviolet light first and then reprogrammed using special equipment. Other disadvantages are a lower performance than ROMs, sensitivity to light and expensive packaging with small quartz windows.

The disadvantages of the EPROM have led to the introduction of the electrically erasable programmable ROM (EEPROM). The EEPROM can be electrically reprogrammed in the system, which eliminates the effort of removing it, erasing it with UV light and reprogramming it using special equipment. EEPROMs allow for low cost packaging and are not sensitive to light. The disadvantages here are the increased cell complexity, when compared with the EPROM, and the limited number of times the EEPROM may be reprogrammed. EEPROMs are only feasible when non-volatility and in-system re-programmability are required.

Based on the previous discussion of different types of memory devices, the advantages of semiconductor memories are clear, but the real edge they have over other types of memory devices is their exceptional performance. Semiconductor memories are the best qualified storage devices to provide for the ever increasing demands of the microprocessor.

1.3 Modeling and testing faults in SRAMs

Depending on the number of operations allowed to be performed simultaneously to SRAMs, a distinction can be made between single-port and multi-port memories. A *single-port (SP)* memory can only be read or written via a single circuit path at a time. A *multi-port (MP)* memory has multiple ports, used to access memory cells simultaneously and independently of each other, subject to certain restrictions; e.g., no cell can be written by more than one port concurrently, in order to prevent inconsistent data values.

1.3.1 Single-port SRAM testing

Tests for SP memories have experienced a long development process. Before 1980, tests had very long test times for a given *fault coverage* (i.e., the number of detected faults divided by the number of total faults); typically of order $O(n^2)$, where n is the size of the memory. Such tests can be classified as the *ad-hoc tests* because of the absence of fault models and proofs. Tests like the *Zero-One test*, the *GALPAT*, and the *Walking 1/0* tests belong to this class [11].

To reduce the test time and improve the fault coverage, test development has been focused on the possible faults which probably can appear in the memory. For that reason, *functional fault models*, which are abstract fault models, have been introduced during the early 1980's. The advantage of these models is that the fault coverage could be proven while the test time usually was of order $O(n)$; i.e., linear with the size of the memory. Some important functional fault models introduced in that time period were the *Stuck-At Fault (SAF)*, the *Address decoder Fault (AF)* [89], the *Coupling Fault (CF)* [49, 59, 87], and the *Neighborhood Pattern Sensitive Fault (NPSF)* model [12, 66]. *March* tests became the dominant type of tests for SAFs, AFs, and CFs [21, 87, 88]; while special linear tests were designed for NPSFs [54, 88].

After the above functional fault models, which were *abstract* fault models not based on real memory design and/or real defects, *Inductive Fault Analysis (IFA)* [48, 78] was introduced. IFA is a systematic procedure to predict the faults in an integrated circuit by injecting spot defects in the simulated geometric representation of the circuit. It allows for the establishment of the fault models based on simulated defects in real designs. In addition, IFA is capable of determining the occurrence probability of each fault model. The result was that new functional fault models were introduced [21]: the *State Coupling Fault* (CF_{st}), the *Data Retention Fault (DRF)*, and the *Stuck Open Fault (SOF)*.

In the early 1990's memories experienced an impressive increase in size, and as a consequence linear tests became not always acceptable. In addition, the use of *embedded* memories had made the testability problem very hard because of the lack of the *controllability* of their inputs and the *observability* of their outputs. Therefore, *built-in-self-test (BIST)* was proposed in order to overcome this problem [43, 44, 53, 69, 71].

BIST has the advantage of solving the access problem of embedded memories, and reducing test requirements in terms of test speed and the number of input/output (I/O) pins. The additional advantage of BIST is that *at-speed testing* (i.e., testing at the maximal clock period) is facilitated allowing for a higher fault coverage, especially for the *dynamic (i.e., speed related) faults*, which are becoming more important with the current high speed memories. The current status of dynamic faults is comparable with the status of functional faults in the early 1980's; i.e., their existence and occurrence probability have not been verified with IFA, or other techniques, while industrial data on their occurrence frequency and on the effectiveness of tests to detect them is not available. *Design for Testability (DFT)* (i.e., make the design manageable) techniques are being implemented to reduce test time and/or to allow for active, rather than passive, tests for cell stability faults such as DRFs [55].

Beside the functional testing, parametric testing has been also developed. A widely used test is the I_{DDQ} *test*, which measures the quiescent power supply current while the circuit is not switching [34, 47, 84]. If for example, a short is present in the circuit, the current may be higher than usual and the fault will be detected. With the I_{DDQ} test defects can be traced which cause no functional faults and which thus can not be detected with functional tests.

1.3.2   Multi-port SRAM testing

MP memories are more complex than SP memories. It therefore stands to reason that tests for MP memories are even more in their infancy stage. In spite of their growing use in telecommunications ASICs and multi-processor systems, little has been published on their fault modeling and testing.

An *ad-hoc test* technique with *no specific fault model* for *dual-port* (i.e., two-port (2P)) memories was described in [50, 68]. The scheme assumes that conventional SP tests are accurate enough to test MP memories; it performs a test with an ascending address on one port and another test with descending address on the other port. Due to the fact that no realistic fault models have been considered, the fault coverage and the effectiveness of this technique to cover faults that can occur in MP memories is very low.

The first effort for *theoretical* fault modeling of specific faults in MP memories was concentrated on *inter-port* faults; i.e, bridges between word lines or bit lines (excluded bridges between bit lines and word lines) belonging to different ports [58, 97, 98, 100]. The inter-port faults can be eliminated by placing V_{dd} of V_{ss} tracks between word lines and bit lines from different ports. In this case, the inter-port fault results in SAFs in some cells, unaddressable cells or cells with multiple addresses. These faults are detectable with conventional SP tests [58]. However, this is a design methodology that requires modification of memory layout, which is very often impossible, especially for memories with large number of ports [97]. Placing V_{dd} of V_{ss} tracks can result in unacceptable silicon overhead.

A Serial BIST circuit technique, generally known as the *shadow write*, was developed at Nortel [58, 97]. Shadow write is a feature of some Nortel's embedded MP memories. When a port is placed in a shadow write mode, its write drivers continuously force a constant value on all bit lines with its word lines internally disabled. Thus if no inter-port fault exists between this port and the port under test, no memory cell will be affected by the shadow write. On the other hand, if a fault does exist, the value read out from the port under test will be affected by the value driven from the shadow write. Although shadow write simplifies the detection of inter-port faults, it requires a modification of the memory design, and also imposes some performance penalties and silicon costs.

For the same fault models (i.e., inter-port faults) as in [58, 97], modified detection approaches have been described in [98, 100]. The proposed scheme in [98] performs a conventional SP test on one port and simultaneously performs an inter-port test on all other ports. The introduced approach in [100] uses two algorithmic steps. In the first step, the modified March C [49] is executed in such a way that read operations are applied simultaneously to all ports; while in the second step a special test, which utilizes simultaneous write operations, is performed.

In [8, 63], a new *theoretical* fault model, called *complex coupling fault* has been developed; this fault model is composed of a set of *idempotent coupling faults (CF_{id}s)* as they may appear in SP memories. However, the individual $CF_{id}s$, of which the complex coupling fault is composed, are too *weak* to sensitize a fault. Their fault effects may be combined when they are activated through different ports simultaneously. This makes the complex coupling fault unique for MP memories. In [8, 63], tests for this fault model have also been presented. Other modified test approach for the same faults have been also proposed. In [99], a pseudo-random test has been described; however, it requires a very long test sequence (approximately 25 times as long as deterministic testing) to obtain an acceptable fault coverage. In [15], a programmable boundary scan technique, using modified tests of those proposed in [8, 63], has been introduced; while in [16] march tests for complex coupling faults in two-port memories, with a time complexity of $\Theta(n^2)$, were developed.

The limitation of the complex coupling fault model lays in three facts. First, the introduced model is only based on weak $CF_{id}s$, while many other fault models may exhibit such weak behavior. Second, the model is not based on any experimental/industrial analysis that can establish its validity. Finally, the introduced tests for such fault models have a test length which is exponentially proportional with the number of ports the MP memory consists of. For instance, for two- and three-port memories, the test length is $19n^2 - 19n$, respectively, $13n^3 - 39n^2 + 26n$, whereby n is the size of the memory [63]. This exponential test length can be reduced to linear by utilizing the topological information of the memory. However, the reduced versions still remain not practical, especially for big memories. For instance, the reduced version of a two-port memory test has a length of $456n$, while that of a three-port memory has a test length of $5148n$ [63].

The state of the art in testing SP memories has resulted in tests which are becoming inadequate for today's high-speed memories because of the following reasons:

- Speed, in terms of insufficient recovery time and set-up time, is becoming a
  dominant failure mode. Adequate fault models and tests still remain to be
  established for this.

- DFT techniques are required in order to guarantee the detection of certain
  time-dependent faults (such as DRFs) within a short test time.

- DFT techniques are required to reduce the test time. This especially applies to
  speed-related faults, which currently are detected by applying tests with many
  variations in addressing sequences and data backgrounds.

- Faults which are of a global nature, such as ground bounce, and faults which
  are neighborhood dependent are not well modeled.

Considering the state of the art in designing fault models and tests for MP memories,
the following can be concluded:

- Adequate fault models remain still to be established. A framework of all possible fault behaviors has to be found. These fault behaviors can be translated into
  functional fault models, which have to be verified through simulation industrial
  level designs.

- Optimal tests have still to be established. That can be done by taking only
  realistic fault models (i.e., verified through simulation) and/or topological information into consideration. Exponential test algorithms are not attractive
  industrially. In order to appreciate the test time consequences, Table 1.3 is
  included. It shows the order of magnitude figures for the required test time,
  assuming a cycle of $10ns$, for MP memories with different sizes and with test
  algorithms with different time complexity. Note that, even for small arrays,
  the test time can become significant!

Table 1.3. Test time as function of the algorithm complexity

	16Kbits	32Kbits	64Kbits
$O(n)$	$163.84\mu s$	$327.68\mu s$	$655.360\mu s$
$O(n^2)$	$2.684s$	$10.737s$	$42.949s$
$O(n^3)$	$12.216h$	$97.734h$	$781.874h$

- A test strategy, in terms of which faults can be detected through a test for
  SP memories versus a test for MP memories, has to be established. This will
  reduce the total test time.

- DFT techniques have to be worked out to further reduce the test time.

- BIST circuits have to be found in order to allow for testing embedded MP memories, and to reduce test requirements.

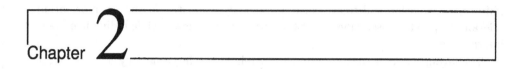

Chapter **2**

Semiconductor memory architecture

2.1 Memory models

2.2 External memory behavior

2.3 Functional memory behavior

2.4 Electrical memory behavior

2.5 Memory process technology

As it is mentioned in Chapter 1, the main purpose of this book is to develop realistic fault models and efficient test algorithms for multi-port SRAMs. It is therefore of interest to understand in detail the structure of such memories as well as the way they operate.

This chapter describes SRAMs in a top down fashion. Each description level is concerned with specific aspects of the SRAM, beginning with its behavior down to the physical buildup of its components. Section 2.1 describes the modeling approach used in this chapter to discuss the structure of the memory. Section 2.2 provides the external SRAM behavior treated as a black box, with only the input and output lines visible. Section 2.3 introduces the functional SRAM model, along with a description of some real SRAM devices. Section 2.4 describes the electrical circuits that make up the functional blocks. Finally, the SRAM cell layouts are treated in Section 2.5.

2.1 Memory models

A system may be described at a number of different levels of abstraction; see Figure
2.1. Each level of abstraction is called a *model* of the system. Models help to sim-
plify the explanation and treatment of systems by explicitly presenting information
relevant only to the discussion about the system at that level, while hiding irrelevant
information.

The layout model is the one most closely related to the actual physical system; it
assumes complete knowledge of the layout of the chip. As we move from right to left
in Figure 2.1, the models become less representative of the physical world and more
related to the way the system behaves, or in other words, less material and more
abstract. A higher level of abstraction contains more explicit information about the
way a system is expected to function and less about its buildup. It is possible to have
a model that contains components from different levels of abstraction; this approach
is called *mixed-level modeling*. With mixed-level modeling, one may focus on low-
level details only in the area of interest in the system, while maintaining high-level
models for the rest of the system.

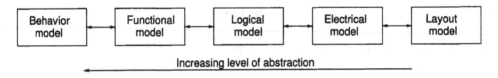

Figure 2.1. Models and levels of abstraction

The modeling levels in Figure 2.1 are explained as follows [88]:

- **The behavioral model:** This is the highest modeling level in the figure and
  is based on the system specifications. At this level, the only information given
  is the relation between input and output signals while treating the system as
  a black box. A model at this level usually makes use of timing diagrams to
  convey information. In this chapter, the behavioral SRAM model is presented
  in Section 2.2.

- **The functional model:** This model distinguishes functions the system needs
  to fulfill in order to operate properly. At this level, the system is divided into
  several interacting subsystems each with a specific function. Each subsystem is
  basically a black box called a *functional block* with its own behavioral model.
  The collective operation of the functional blocks result in the proper operation
  of the system as a whole. This model is given in Section 2.3.

- **The logical model:** This model is based on the logic gate representation of the system. At this level, simple boolean relations are used to establish the desired system functionality. It is not the custom to model memories exclusively using logic gates, whereas logic gates are sometimes present in models of a higher or lower levels of abstraction to serve special purposes. Therefore, no exclusive SRAM logical model is given in this chapter.

- **The electrical model:** This model is based on the basic electrical components that build up the system. In semiconductor memories, the components are mostly transistors, resistors and capacitors. At this level, we are not only concerned with the logical interpretation of an electrical signal but also the actual electrical values of it. Since this research is partially concerned with experimental analysis at the electrical level of (multi-port) SRAM circuits, this memory model is treated in more depth. This model is given in Section 2.4.

- **The layout model:** This is the lowest modeling level available. It is directly related to the actual physical implementation of the system. At this level, all aspects of the system are taken into consideration, even the geometrical configuration, such as distances and thickness of lines, matters. For this reason, this model is also called the geometrical model. The data representing this model are rarely reported in the literature. Section 2.5 discusses briefly the representation of SRAM cells according to layout model.

Paying a closer attention to the behavioral and the functional models reveals that there is a strong correspondence between the two. In fact, the behavioral model can be treated as a special case of the functional model, with the condition that only one function is presented, namely the function of the system itself. Therefore, some authors prefer to classify both modeling schemes as special cases of a more general model called the *structural model* [3]. The structural model describes a system as a number of interconnected functional blocks. According to this definition, a behavioral model is a structural model with only one function, while a functional model is a structural model with more than one interconnected function.

2.2 External memory behavior

The most general model of a memory is a box with inputs and outputs; see Figure 2.2. As can be seen from the figure, a memory receives address, control, and data-in values from the exterior via the inputs, and produces data-out values via the outputs. The inputs consist of C controls, N address lines, and B input data lines, whereby B is the *word-width* of the memory. The outputs consist only of a $B-$bit data-word. Usually the data-in lines and the data-out lines are combined to form bidirectional data lines, thus reducing the number of pins of the chip.

Figure 2.2. General model of a SRAM

The model of Figure 2.2 can be further refined into two parts, the so called *two-dimensional memory model* [65]; see Figure 2.3. The two parts of this memory model are:

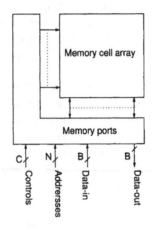

Figure 2.3. Two-dimensional SRAM model

- The *memory cell array* part: this part represents the memory cell array. Every memory cell can store only one bit of information. In addition, there is always exactly one memory cell array in a memory.

- The *memory ports* part: this part represents the interface between the external world of the memory and the memory cell array. A memory can have p memory ports, whereby $p \geq 1$.

Note that at this level of abstraction, nothing is supposed to be known about the internal structure of the SRAM; it contains the least information any memory sheet must provide to describe the working of the circuit. Usually, data sheets use *timing diagrams*, which specify the minimum and the maximum timing requirements for the circuit.

2.3   Functional memory behavior

As mentioned in Chapter 1, depending on the number of operations allowed to be performed simultaneously to SRAMs, a distinction can be made between single- and multi-port memories (i.e., SRAMs). A *single-port (SP)* memory is one which can only be read or written via a single circuit path at time. A *multi-port (MP)* memory is a memory having multiple ports that are to be used to access memory cells simultaneously and independently of each other. In this section, functional model of SP and MP memories will be discussed separately.

2.3.1   Functional SP SRAM model

A typical SP memory consists of a memory cell array, two address decoders, read/write circuits, data flow and control circuits; see Figure 2.4. The memory chip is connected to other devices through address lines, data lines, and control lines (i.e., read/write line, chip enable line, and power lines).

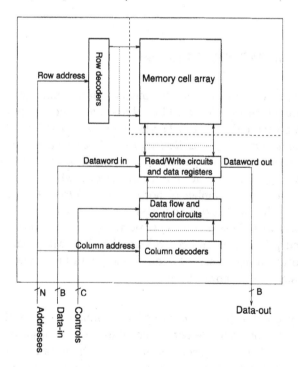

Figure 2.4. Functional model of a SP memory

The memory cell array is the most basic part of the memory. It consists of n cells, which are organized as an array of R rows and C columns. The number of rows can be any integer, but the number of columns is restricted: there is always an integer number of memory *words* in one row (i.e., $C \bmod B = 0$). Note that the memory cell array has a capacity of $R \times C$ bits.

The addresses are divided into high- and low-order bits. The high-order bits are connected to the row decoder which selects an appropriate row (*Word Line*) in the memory cell array, while the low-order bits are connected to the column decoder which selects the required columns (*Bit Lines*). The number of columns selected is B, which determines how many bits can be accessed during a read or a write operation.

To read the desired memory cells, appropriate row and column select lines must be activated. The content of the selected cells are amplified by the read circuits, loaded into the data registers and presented on the data-out lines. Conversely, during a write operation, the data on the data-in lines is loaded into the data registers and written into the selected cells through the write circuits. Usually the data-in and data-out lines are combined to form bidirectional data lines, thus reducing the number of pins of the chip.

2.3.2   Functional MP SRAM model

To transform the functional model of a SP memory into a functional model of a MP memory, ports are added by adding address lines, data lines, chip enable lines and decoders for each port; see Figure 2.5. Furthermore, each port may have the capability to be a *read-only port (Pro)*, a *write-only port (Pwo)*, or a *read-write port (Prw)*. A read-only port has only the capability of reading a memory word; a write-only port has only the capability of writing a memory word; while a read-write port has both capabilities. The total number of ports is $p = (\#Pro) + (\#Pwo) + (\#Prw)$ (e.g., $\#Pro$ denotes the number of *Pro* ports) . A MP memory consists of a common memory cell array, two address decoders per port, p (number of ports) of read and/or write circuits, and p data flow and control circuits. The MP memory is connected to the outside world through p sets of address lines, data lines, read/write lines, and chip enables lines as well as power lines.

The MP memory operations are performed in a similar way as those of the SP memory, except that now multiple ports control up to p simultaneous operations to one or more words. However, various bizarre interactions can occur as a result of the simultaneous multi-port operations; e.g., a word may be read and written at the same time, a word may be written by more than one port concurrently such that inconsistent data values may result. Port *priority-style* may be used [23]: several ports can simultaneously access the memory to write different data words to the same cells, but only the port with the highest priority succeeds.

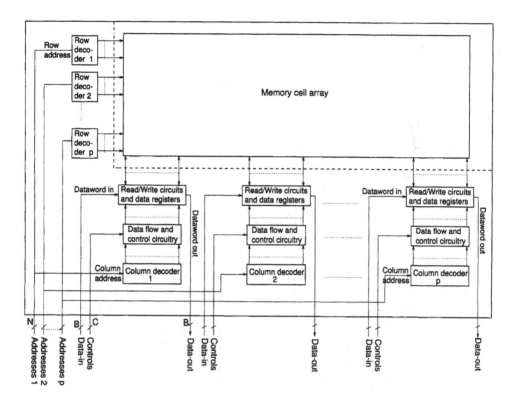

Figure 2.5. Functional model of MP memory

2.4 Electrical memory behavior

This section presents the electrical structures that makes up the functional blocks presented in the previous section for SP and MP memories.

2.4.1 Electrical structure for SP SRAMs

The blocks of the functional model for SP memories presented in Figure 2.4 will be opened such that the electrical properties will become visible. This will be done for memory cells, the address decoders, and the read/write circuits.

Memory cells

The memory cell is the most basic part of the memory. Its design depends on various factors, such as the memory application and the implementation style. For a SP SRAM, the memory cell is a *bistable* circuit, being driven into one of two states. After removing the driving stimulus, the circuit retains its state. An SRAM cell can have several configurations. Figure 2.6 shows the generalized SRAM cell, and three possible configurations.

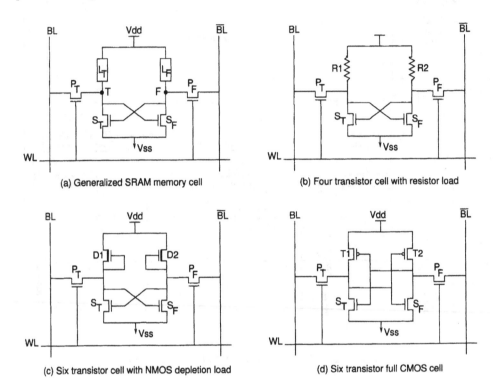

Figure 2.6. Generalized SRAM cell, and various configurations of SRAM cells

As can seen in Figure 2.6(a), the SRAM cell consists of two *load elements* (L_T and L_F), two *storage elements* (S_T and S_F), and two *pass transistors* (P_T and P_F). Transistor S_T forms an inverter together with the load element L_T. This inverter is cross-coupled with the inverter formed by the transistor S_F and the load element L_F; therefore, forming a *latch*. This latch can be accessed for read and write operations, via the pass transistors P_T and P_F.

Data can be written into the cell by driving the lines BL and $\overline{BL}$ with data with complementary values, and thereafter driving the *word line (WL)* high. The cell will be forced to the state presented on BL and $\overline{BL}$, since the two lines are driven with more force than the force with which the cell retains its information. To read data from a cell, generally, first both lines BL and $\overline{BL}$ are *precharged* to a high level, after which the desired WL is driven high. At that time the data in the cell will pull one of the two bit lines low. This difference signal on the BL and $\overline{BL}$ lines is amplified by the read circuit (see Figure 2.4), and read out through the data register. It should be noted that reading an SRAM cell is a non-destructive process; i.e., after the read operation the logic state of the cell remains the same.

The load devices may consist of polysilicon resistors, either enhancement or deple-tion mode transistors, or PMOS transistors. Figure 2.6(b) shows the SRAM cell with polysilicon load devices. This cell requires less silicon area than the two other configurations. However, it has a higher current when it is not being accessed, since a small amount of current always flows through the resistor. When the load element is a PMOS transistor (see Figure 2.6(d)), then the resulting CMOS cell has essen-tially no current drain through the cell, except when it is switching because either the NMOS or the PMOS transistor is always off. The disadvantage of the CMOS cell is that it requires more processing steps because of the presence of NMOS and PMOS transistors. Figure 2.6(c) shows a six-device SRAM cell using depletion mode load transistors. It should be noted that the depletion mode transistor can also be replaced with an enhancement mode transistor, but the depletion load is normally used since it has better switching performance, a higher impedance, and is relatively insensitive to power supply variations [67].

Address decoders

Address decoders are used to access particular cells in the memory cell array. In order to reduce the size of the decoders and the length of the word and bit lines, two dimensional addressing schemes are used within the chip, demanding a row decoder for the word lines and a column decoder for the bit lines.

A row decoder is needed to select one row out of the set of rows in the memory array. Figure 2.7 shows two basic *static row decoders*, namely a *PMOS-load decoder* [72] and a *CMOS decoder*. The inputs of the decoders are the address bits A_0 through A_{k-1} or their complements, while the output is the word line. When the row decoder selects a word line, all cells along that word line are activated and put their data on the bit lines. Note that the address lines are connected only to the NMOS transistors in a PMOS-load decoder; while they are connected to both PMOS and NMOS transistors in a CMOS decoder. Therefore, the address load capacitance caused by the gates in a PMOS-load decoder is almost half of that in a CMOS decoder. This implies that a smaller delay time can be obtained by using a PMOS-load decoder. The CMOS decoder has the advantage of drawing no static current, but as it requires an equal number of PMOS and NMOS transistors, it occupies a larger area.

Dynamic or *clocked decoders* are also used to decode word lines. Figure 2.8 shows two dynamic row decoders. Generally, such decoders combine compact layout with zero static current consumption; power is dissipated in the selected decoder only during the brief period of an address transition. The decoder of Figure 2.8(b) proposed by

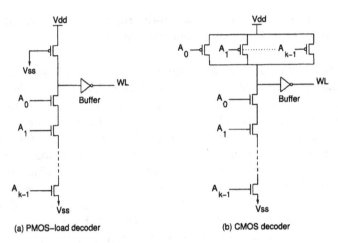

(a) PMOS–load decoder   (b) CMOS decoder

Figure 2.7. Static row decoders

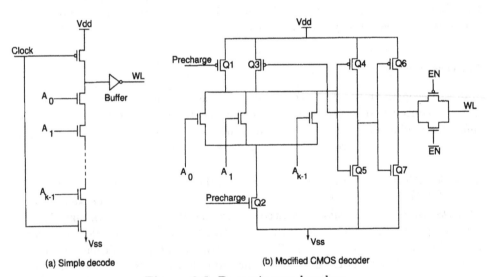

(a) Simple decode   (b) Modified CMOS decoder

Figure 2.8. Dynamic row decoders

[52] operates as follows: in the precharge phase, the transistor $Q1$ is turned on to precharge the common line connected to the address decoding transistors. If all the address bits, A_0, ..., A_{k-1} are zeros, transistor $Q6$ drives the WL line to '1'. The signal EN enables the transmission gate such that the decoder selects the word line only after all address lines are stable.

A column decoder selects B bit line(s) (or bit line pairs) out of the set of bit lines (or bit line pairs) of the selected row. Depending on the memory application, different

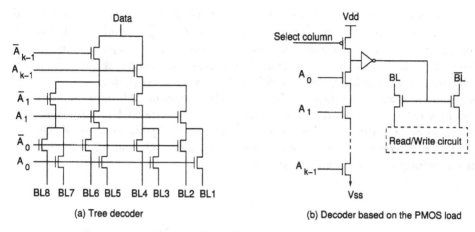

(a) Tree decoder   (b) Decoder based on the PMOS load

Figure 2.9. Column decoders

column decoders are designed. Figure 2.9(a) shows a *tree decoder*, which is desirable for *single ended* memories; i.e., memories which use only a single bit line for read and write operations. This circuit has the advantage of being simple, however it operates slowly. Figure 2.9(b) shows another column decoder [56], which is based on the PMOS-load decoder of Figure 2.7. The output of the decoder goes to an inverter, where the output signal is amplified, after which it moves on to the column switch MOS transistors. This circuit has the advantage of being compact.

Read/Write circuitry

Once a particular single or pair of bit lines have been selected, depending on the SP memory cell structure, a circuitry is required to write or to read the cells. Typical write circuits are shown in Figure 2.10. Circuit (*a*) consists of a pair of inverters and a pass gate with a write control input signal; while circuit (*b*) consists of a pair of NAND gates. The to be written data on 'data in' line is presented on BL and $\overline{BL}$.

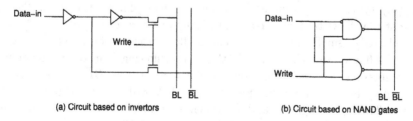

(a) Circuit based on invertors   (b) Circuit based on NAND gates

Figure 2.10. Memory write circuitries

The read circuitry is more complex than the write circuitry and depends on the type of the memory cells to be read, namely *single ended* or *differential*. In addition, it can be based on a *voltage mode* or a *current mode* signal transporting technique. Figure 2.11 shows two voltage mode sense amplifiers, namely a single ended PMOS differential sense amplifier, and a double-ended PMOS cross-coupled amplifier [73]. In circuit (a), when the data on BL is '1', the transistor $M1$ turns on, and the transistor $Q2$ drives the Out line to '1'; while when the data on BL is '0', transistor $M2$ turns on, and drives the Out line to '0'. In circuit (b), the voltages of the Out and $\overline{Out}$ lines control the gates of the PMOS transistors $Q1$ and $Q2$ such that the output voltage transitions are accelerated.

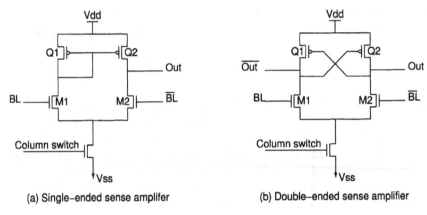

(a) Single–ended sense amplifer   (b) Double–ended sense amplifier

Figure 2.11. Voltage mode sense amplifiers

Current mode sense amplifiers operate generally faster than voltage mode ones. Figure 2.12 shows two current mode sense amplifies: a *double-ended current-mirror* amplifier [74] and a *hybrid current sense* amplifier [17].

The double-ended current-mirror amplifier uses a bias voltage generator to provide an appropriate voltage to the PMOS transistors P1 though P4, so that they operate near the saturation region; see Figure 2.12(a). When no cell is selected, the current (say I_0) that flows from each bit line to the amplifier is the same. Therefore the current that flows through P1-P4 in the amplifier is also the same $(I_0/2)$. When a cell is selected, a small amount of current ΔI flows from one bit line to the cell. For example, ΔI flows from BL to the cell, the current flowing from BL to the amplifier will be reduced to $I_0 - \Delta I$, while the current flowing from $\overline{BL}$ will remain to I_0. Consequently, the current flowing through P1 and P3 will be reduced to $(I_0 - \Delta I)/2$, while that flowing through P2 and P4 remains at $I_0/2$. The current which flows through M1 is $I_0/2$ because the current flowing through P2 and M2 is the same, and M1 is the current mirror of M2. Therefore the load capacitance of output line Out will be discharged by M1, which draws more current than P1 provides. In almost the same manner, the load capacitance of $\overline{Out}$ will be charged up by P4. In this way, the voltage of *Out* drops while the voltage of $\overline{Out}$ rises, and the voltage swing is obtained between the two data output lines.

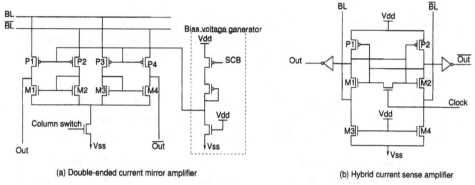

(a) Double-ended current mirror amplifier   (b) Hybrid current sense amplifier

Figure 2.12. Current mode sense amplifiers

The hybrid current sense amplifier (see Figure 2.12(b)) operates in two phases: equalization and sensing phase. During the equalization phase, the clock signal is high to precharge the output nodes to equal potentials. During the sensing phase, a particular memory cell is selected and the cell's node with low level draws a current from the corresponding bit line. The differential current signal then appears at *BL* and $\overline{Out}$. Once the clock signal is low, the differential current flows through M1 and M2, and charging the small equivalent capacitances at the drains of M1 and M2. A small differential voltage will appear across the two drains and be amplified to a CMOS level voltage by the positive feedback effect of the cross coupled circuit.

2.4.2   Electrical structure for MP SRAMs

Describing the electrical model for MP memories will be done in the same way as is done for SP memories; i.e., the blocks of the functional model of Figure 2.5 will be opened such that the electrical properties become visible.

A MP memory consists of p duplicated ports and a common memory cell array. The circuits of each port in the memory port part for MP memory can be the same as these of the SP memory. That means that the address decoders and sense amplifiers discussed in the previous section can also be used for MP memories. However, the memory cell array has to be modified in order to support multiple accesses. In the following, different MP memory cell structures will be described. In addition, two sense amplifiers, special for MP memories, will be discussed.

MP memory cell structures

Depending on the type of port designs the MP memory cells consist of, they can be classified into three classes:

- cells having *shared* read/write ports: all ports of these cells are of the type Prw; that means that each port can be used to read or to write a cell. Moreover, with these cells the same bit line (or bit line pair) is used for reading or writing data.

- cells having *separate* read and write ports: all ports of these cells are of the types Pro or Pwo. The read and write ports have separate paths; i.e., some bit lines are only used for reading data, and others only for writing data.

- cells having *mixed* read and write ports: these cells can have Prw ports, as well as Pro ports and Pwo ports. The operation can be done via a single path as well as via separate paths; i.e., some bit lines are only used for reading or only for writing, and some bit lines are used for both reading and writing.

In the following, each class will be discussed separately.

Shared read/write MP cells: Three shared read/write MP cells with two Prw are shown in Figure 2.13. Figure 2.13(a) shows a *differential access* MP cell [40]. This cell includes the general form for parallel expansion of independent differential ports. Each port has bit lines BL and $\overline{BL}$ such that a differential sense amplifier can be used. In addition, for each port two pass transistors are added to the basic four-transistors memory cell. Because of the complex word and bit line interconnections and the two extra pass transistors for each port, this cell requires more area. However, its performance is better than that of a single-ended cell.

The circuit of Figure 2.13(b) shows a *single-ended* MP cell [80]. In this case, for each read/write port only one pass transistor is added to the basic four-transistor cell. Therefore, the area of the memory chip is minimized. However, its performance is lower than that of a cell with differential access cell, since differential sense amplifiers can not be used.

Figure 2.13(c) shows a *twin-port* cell [64] which uses complementary access devices. In addition, the symmetric port structure, that matches the cell symmetry, allows for an effective use of the CMOS cell since each port has the potential of accessing a cell without interference from activities at the other port, even if addressing the same cell. The read operation is based on the complementary precharging, i.e., bit line BL_{NH} is precharged high and the bit line BL_{PL} is precharged low. If the port 1 (P-channel pass transistor) is read and the stored data is '1', device $M6$ will conduct, charging the bit line BL_{PL} from the V_{SS} to a positive level. But if the stored data is '0', device $M6$ will not conduct, as the gate, drain and source are all at V_{SS} (passive reading). The read operation by port 2 is done similar as by port

1. Note that the circuit of Figure 2.13(c) can conceptually be extended in such way that it will be differential.

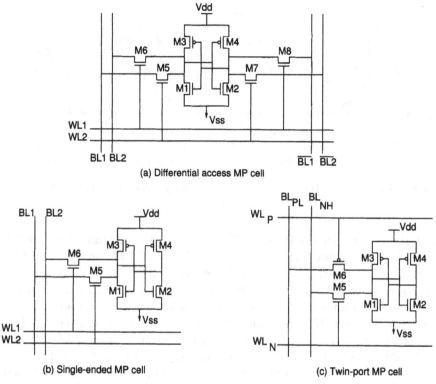

Figure 2.13. Common read/write MP memory cells with Prw=2

Separate read and write MP cells: Figure 2.14 shows two implementations of MP cells with separate read and write ports. The circuit of Figure 2.14(a) [83] shows a four port memory cell, with two Pro and two Pwo. A single bit line is used for reading data, and another single bit line is used for writing data. Usually this cell has separate read and write ports; i.e., no read/write port. It has the best electrical characteristics, since the addition of transistors $M7$ and $M8$ allows for a read operation to have a very small load on the cell. However, it occupies a large area.

Figure 2.14(b) [46] gives another implementation of a MP cell with separate read and write ports. This circuit has four ports: two Pro and two Pwo. A single bit line scheme is used for both read and write operations. The circuit that drives the bit line during the read operation, i.e., the inverter formed by the transistors $M7$ and $M8$, is unidirectional. Regardless of the bit line capacitance of the number of the ports enabled, the voltage change on the node N due to the capacitive coupling with a bit line, cannot switch the memory cell; this is because the false node of the cell is isolated from the bit lines used for read operations.

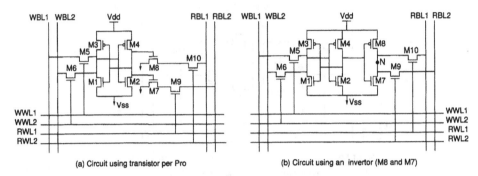

Figure 2.14. Separate read and write MP memory cells with Pro=2, and Pwo=2

Mixed read and write MP cells: Two different implementations of mixed MP cells are shown in Figure 2.15. The circuit of Figure 2.15(a) consists of a MP cell with two Pro and a single Prw [80]. Each port has a single bit line for reading data. Therefore, single-ended sense amplifiers are needed. The write operation is done by driving both word lines $WL1$ and $WL2$ high, and putting $DATA$ and $\overline{DATA}$ on the bit lines $BL1$ and $BL2$. Note that this cell combines the read scheme of Figure 2.13(a) for bit line count reduction, and the write scheme of Figure 2.13(c) for a fast write. Another implementation of such memory cell, is given in Figure 2.15(b) [96]. This lower power circuit consists of a Prw which is based on voltage mode access (bit lines $BL1$ and $\overline{BL1}$), and a Pro which is based on current mode access (bit line $BL2$).

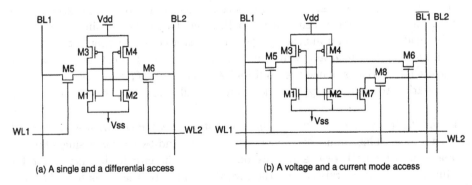

Figure 2.15. Mixed ports MP memory cells

Sense amplifiers for MP memories

In this subsection, two sense amplifiers, which are especially designed for MP memories, will be discussed. The first one is a voltage mode sense amplifier based on a *dummy cell* [81], and the second one is a current mode sense amplifier based on *current direction* sensing [36].

Figure 2.16 shows the voltage mode sense amplifier. This circuit uses a dummy cell to generate a reference voltage for a differential sense amplifier in order to achieve high speed sensing for single-ended MP memories. When the internal clock goes low, the read bit line RBL and dummy bit line DBL are precharged. When the clock goes high, the dummy cell and the MP memory cell are simultaneously selected. the differential signals are fed to the pre-amplifier, that quickly shifts these differential signals to the level which causes the main amplifier to operate most sensitively. Finally the shifted signals are sensed by the differential amplifier.

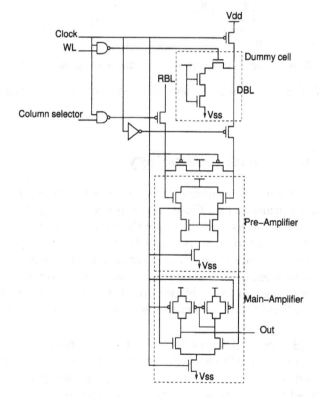

Figure 2.16. Voltage mode sense amplifier based on dummy cell

Figure 2.17 shows the current mode sense amplifier based on current direction sensing. It reduces the access time of a single-ended MP memory cell without the need for a reference voltage. The voltage at the current direction sense circuit's input node referred to as INT, is maintained at an intermediate voltage level. When the input is '0', the current will be pulled from the INT; while when it is '1', the current will be pushed to the INT. The current direction sense circuit reacts to these varying stimuli by producing respectively different differential outputs to the second amplifier, which in turn amplifies these respective voltage differentials to the CMOS level.

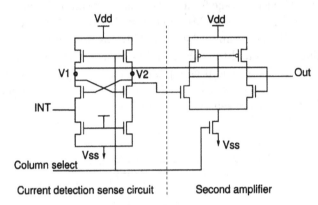

Figure 2.17. Current mode sense amplifier based on current direction sensing

2.5   Memory process technology

The last step in designing or analyzing memories is the description of their layout or their physical implementation. The design step from the electrical to the layout level is not as small as it may seem. Despite the great advancement in the field of computer aided design (CAD), memory design is still a labor-intensive activity. This is due to competition pressure to keep the chip area, and thus the price, minimal. On the other hand, simulation tools are being used extensively to verify the operation of the memory design. The real layout of memories is rarely reported in the literature. In the following, a basic MOS memory process will be discussed. For more details see e.g., [95].

The basic MOS integrated circuit process begins with the growth of large cylindrical single crystals of pure silicon which are sliced into disks called 'wafers'. These wafers are thinned to about $400\mu m$ after processing, and are generally between 20 and $30cm$ in diameter. Silicon is used as the base material because as a semiconductor it has a resistivity that can be changed by introducing special impurity atoms in the crystal. The memory circuit is constructed on the surface of the wafer using series of successive steps to form patterns alternated with diffusion or ion implantation of various chemical dopants into the openings of the pattern. The patterning uses a number of masks to define the patterns in a process called photolithography. Figure 2.18 shows the most interesting steps for a MOS process, and are explained below.

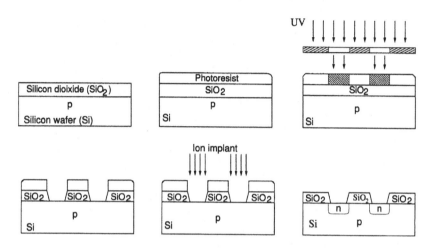

Figure 2.18. Basic MOS process

The process starts with the wafer; the silicon can be doped with material that gives rise to positive or negative charge carriers, resulting in making the wafer a positively or negatively doped semiconductor. In the figure, it is assumed that the silicon is doped with positive (p) charge carriers. Then, a thermal oxidation of the silicon surface takes place, by which an oxide layer (SiO_2) is created on the silicon. The oxide is then covered with layers of photoresist, which is a light sensitive polymer. If the photoresist is exposed to UV light, it becomes soluble in certain solutions. This step is followed by setting a patterned mask over the wafer, through which UV light passes and transforms the exposed photoresist areas into a soluble material. The mask will define what will eventually be sources, drains, channels and diffused crosses. Then, the exposed areas are removed using, e.g., a chemical etch. Dopant atoms are then introduced into the silicon; this is done either in a high temperature gas environment or with an ion beam accelerator which implants ions in the patters defined on the silicon surface of the wafer. The implantation causes the silicon to act either as N-type or P-type doped. Finally, the remaining photoresist can be stripped from the SiO_2 surface using special solvents. This process may continue to produce quite complicated configurations of patterned layers on the silicon wafer.

Another step, which is not shown in the figure, would involve depositing a doped amorphous polysilicon or metal on the layer surface. This is then patterned to form interconnects between the elements on the wafer surface.

Figure 2.18

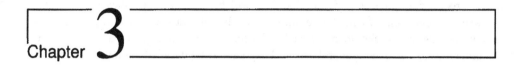

Chapter **3**

Space of memory faults

3.1 Concept of fault primitive

3.2 Classification of fault primitives

3.3 Single-port faults

3.4 Two-port faults

In Chapter 2, different SRAM models have been discussed; e.g., the functional model, which is the collection of the functional specifications of the memory together with the internal structure of its subsystems. In general, the functional model of a memory depends on its specific implementation. However, for test purposes a so called 'reduced functional memory model' is used that only consists of three subsystems: the address decoder, the memory cell array and the read/write logic. Since the vast majority of mainstream memory devices contains these three subsystems, the reduced functional fault model is, to a large extent, independent of specific memory implementations.

This chapter is concerned with defining the functional fault models used in this book. The concept of 'fault primitives' will be used to determine a general framework with which the space of functional fault models can be studied and analyzed. Section 3.1 defines the concept of fault primitives and functional fault models. Section 3.2 classifies the faults in SRAMs and shows the scope of faults targeted. Section 3.3 presents a complete set of single-port faults, while Section 3.4 introduces a complete set of two-port faults.

3.1   Concept of fault primitive

Intuitively, a functional fault model is defined as a description of the failure of the memory to fulfill its functional specifications. This definition of a fault model is not a precise one since it does not indicate which functional specifications should be taken into account. Still, the definition specifies the intuitive meaning of a fault model and the way it should be viewed. The term 'functional specifications' should be understood in a rather general sense. They should be detailed enough to describe the contents of individual memory cells.

By performing a number of memory operations and observing the behavior of any component functionally modeled in the memory, functional faults can be defined as the deviation of the observed behavior from the specified one under the performed operation(s). Therefore, the two basic ingredients to any fault model are:

1. A list of performed memory operations.

2. A list of corresponding deviations in the observed behavior from the expected one.

Any list of performed operations on the memory is called an *operation sequence.* An operation sequence that results in a difference between the observed and the expected memory behavior is called a *sensitizing operation sequence (SOS).* The observed memory behavior that deviates from the expected one is called a *faulty behavior.* When inspecting the memory for possible faulty behavior, not all the functional specifications are taken into account and compared with the actual memory behavior. Rather, a very limited subset of functional parameters is selected as most relevant to describe the faulty behavior of the memory. Throughout the 1980s and during the first half of the 1990s, the only functional parameter considered relevant to the faulty behavior was the stored logic value in the cell. Recently, another functional parameter, the output value of a read operation, was also considered to be relevant to describe the faulty behavior.

Thus in order to specify a certain fault, one has to specify the SOS, together with the corresponding faulty behavior. This combination for a single fault behavior is called a *Fault Primitive (FP)* [93], and is denoted as $< S/F/R >$. S describes the SOS that sensitizes the fault, F describes the value or the behavior of the faulty cell (e.g., the cell flips from 0 to 1), while R describes the logic output level of a read operation (e.g., 0).

The concept of a FP allows for establishing a complete framework of all memory faults, since for all allowed operation sequences in the memory, one can derive all possible faulty behaviors. In addition, the concept of a FP makes it possible to give a precise definition of a *functional fault model (FFM)* as it has to be understood for memory devices [93]:

A **functional fault model** is a non-empty set of fault primitives

This definition of a FFM still depends on the selected functional parameters to be observed in the FPs. Yet, this dependence is now precisely known once the FPs are defined. Since a fault model is defined as a set of FPs, it is expected that FFMs would inherit the properties of FPs. For example, if a FFM is defined as a collection of single cell FPs, then the FFM is a single cell fault. If a FFM is defined as a collection of 2-operation (i.e., the SOS consist of two sequential operations) FPs, then the FFM is also called a 2-operation fault.

The situation becomes more complicated if a FFM consists of FPs classified into inconsistent classes (e.g., single cell and two-cell FPs). In this case, the FFM is not described by a single term but by the classes of its constituent FPs. Therefore, a FFM that consists of single cell and two-cell FPs, for example, is described as a single and two-cell FFM.

3.2 Classification of fault primitives

Figure 3.1 shows the different classifications of the FPs. They can be classified based on:

1. the number of *sequential* operations required in the SOS, into *static* and *dynamic* faults.

2. the way the FPs manifest themselves, into *simple* and *linked* faults.

3. the number of *simultaneous* operations required in the SOS, into *single-port* and *multi-port* faults.

4. the number of different cells the FPs do involve, into *single-cell* and *multi-cell* faults.

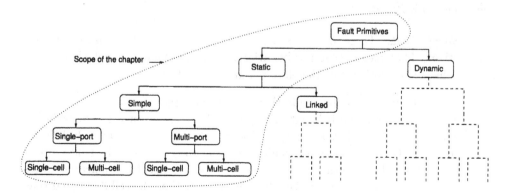

Figure 3.1. Summary of fault primitive classification

It is important to note that the four ways of classifying fault primitives are independent since their definition is based on independent factors of the SOS; see Figure 3.1. As a result, a dynamic fault primitive can be single-port or multi-port, single-cell or multi-cell. The same is true for linked faults; they can be static or dynamic, and each of them can be single-port or multi-port, single-cell or multi-cell.

In this chapter, we will focus on *static, simple* faults; including, single-port versus multi-port, and single-cell versus multi-cell; see Figure 3.1. From here on, the term 'fault' or 'fault primitive' refers to a 'static, simple fault primitive'. In the rest of this section, the FP classification is discussed in more detail.

3.2.1   Static versus dynamic faults

Let $\#O$ be defined as the number of different operations performed *sequentially* in a SOS. For example, if a single read operation applied to a certain cell causes the same cell to flip, then $\#O = 1$. Depending on $\#O$, FPs can be divided into *static* and *dynamic* faults:

- *Static faults*: These are FPs sensitized by performing *at the most* one operation; that is $\#O \leq 1$. For example, the state of the cell is always stuck at *one* ($\#O = 0$), a read operation to a certain cell causes the same cell to flip ($\#O = 1$), etc.

- *Dynamic faults*: These are FPs that can only be sensitized by performing more than one operation sequentially; that is $\#O > 1$. Depending on $\#O$, a further classification can be made between *2-operation dynamic FPs* whereby $\#O = 2$, *3-operation dynamic FPs* whereby $\#O = 3$, etc.

3.2.2   Simple versus linked faults

Depending on the way FPs manifest themselves, they can be divided into *simple faults* and *linked faults*.

- *Simple faults*: These are faults which can not influence the behavior of each other. That means that the behavior of a simple fault can not change the behavior of another one; therefore *masking* can not occur.

- *Linked faults*: These are faults that do influence the behavior of each other. That means that the behavior of a certain fault can change the behavior of another one such that *masking* can occur [66, 88]. Note that linked faults consist of two or more simple faults. In order to get more insight into linked faults, the following example will be given. Assume that the application of an operation to a cell c_1 will cause a fault in a cell c_v (i.e., the cell flips); and that the application of an operation to a cell c_2 will cause a fault in the same cell c_v, but with a fault effect opposite to that caused by cell c_1. If now first an operation is applied to cell c_1, and thereafter to cell c_2, then the net result is that the fault effect of cell c_1 is masked by the fault effect of cell c_2; i.e., no fault effect is then visible in cell c_v.

3.2.3   Single-port versus multi-port faults

Let $\#P$ be defined as the number of ports required *simultaneously* to apply a SOS. For example, if a single read operation applied to cell c_1 causes the same cell to flip, then $\#P = 1$; if two *simultaneous* read operations applied to the cell c_1 cause the same cell to flip, then $\#P = 2$. Depending on $\#P$, FPs can be divided into *single-port* faults, and *multi-port* faults.

- *Single-port faults (1PFs)*: These are FPs that require *at the most* one port in order to be sensitized; that is $\#P \leq 1$. Note that single-port faults can be sensitized in single-port memories as well as in multi-port memories.

- *Multi-port faults (pPFs)*: These are FPs that can be only sensitized by performing two or more simultaneous operations via the different ports. Depending on $\#P$, the multi-port faults can be further divided into:

  - *Two-port faults (2PFs)*: These are FPs that can be only sensitized by performing two simultaneous operations via two different ports; that is $\#P = 2$. Note that 2PFs can be sensitized in any multi-port memory with $p \geq 2$ (p denotes the number of ports).
  - *Three-port faults (3PFs)*: These are FPs that can only be sensitized by performing three simultaneous operations via three different ports; that is $\#P = 3$. Note that 3PFs can be sensitized in any multi-port memory with $p \geq 3$.
  - etc.

3.2.4   Single-cell versus multi-cell faults

Let $\#C$ be defined as the number of different cells accessed during a SOS. For example, if the operation sequence consists only of a single read operation applied to a single cell, then $\#C = 1$; if the operation sequence consists of two single read operations applied sequentially to two different cells, then $\#C = 2$; etc. Depending on $\#C$, FPs can be divided into *single cell faults* and *multi-cell faults* (i.e., *coupling faults*).

- *Single-cell faults*: These are FPs involving only a single cell. They have the property that the cell used for sensitizing the fault (by applying the SOS) is the *same* as where the fault appears.

- *Coupling faults*: These are FPs that involve more than one cell; they have the property that the cell(s) which sensitizes (or contribute for sensitizing) the fault (e.g., by applying the SOS) is *different* from the cell where the fault appears. Depending on $\#C$, this class can be further divided into *two-coupling fault primitives* whereby $\#C = 2$, *3-coupling fault primitives* whereby $\#C = 3$, etc.

3.3   Single-port faults

Single-port faults occur in single-port memories as well as in multi-port memories, and can be divided into single-cell FPs and multi-cell FPs. Single-cell FPs are FPs involving a single cell; while multi-cell FPs are FPs involving more than one cell. For multi-cell FPs, we will restrict our analysis to two-cell FPs (i.e., two-coupling FPs), because they are considered an important class in single-port SRAM faults.

Figure 3.2 shows the two classes considered for 1PFs. *Single-port faults involving a single cell (1PF1s)* have the property that the cell used for sensitizing the fault is the same cell as where the fault appears. On the other hand, *single-port faults involving two cells (1PF2s)* are divided into three types, depending on the cell to which the sensitizing operation is applied:

1. The $1PF2_s$: It has the the property that the *state* of the *aggressor cell (a-cell c_a)*, rather than an operation applied to c_a, sensitizes a fault in the *the victim cell (v-cell c_v)*. Note that no operation is required in that case; the subscript 's' in the notation $1PF2_s$ stands for 'state'.

2. The $1PF2_a$: It has the the property that the application of a single-port operation (solid arrow in the figure) to the *a-cell* sensitizes a fault in the v-cell.

3. The $1PF2_v$: It has the the property that the application of a single-port operation to the *v-cell*, with the a-cell in a certain state, sensitizes a fault in the v-cell.

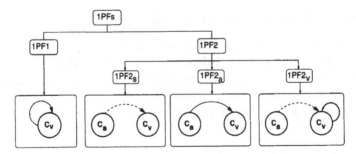

Figure 3.2. Classification of 1PFs

In the rest of this section, the domain of all possible 1PFs and 2PFs will be presented. For each class, first the complete list of FPs is given, thereafter the list will be compiled into FFMs.

3.3.1   Single-cell fault primitives

Before listing the possible single-cell FPs (1PF1s), a precise compact notation, which will prevent ambiguities and misunderstandings, will be introduced.

$< S/F/R >$ (or $< S/F/R >_v$): denotes an FP involving a single-cell (i.e., 1PF1s); the cell c_v (victim cell) used to sensitize a fault is the same cell as where the fault appears.

S describes the value/operation *sensitizing* the fault; $S \in \{0, 1, 0w0, 1w1, 0w1, 1w0, 0r0, 1r1\}$, whereby 0 (1) denotes a *zero* (*one*) value, $0w0$ ($1w1$) denotes a write 0 (1) operation to a cell which contains a 0 (1), $0w1$ ($1w0$) denotes an up (down) transition write operation, and $0r0$ ($1r1$) denotes a read 0 (1) operation. If the fault effect of S appears after a time T, then the sensitizing operation is given as S_T.

F describes the value of the *faulty* cell (v-cell); $F \in \{0, 1, ?\}$, where 1 (0) denotes that the state of the victim cell becomes 1 (0) (or remains in 1 (0)) due to a certain sensitizing operation/value; and ? denotes an *undefined* state of the cell (e.g., the voltage of the true and the false node of the cell are almost the same).

R describes the logical value which appears at the output of the SRAM if the sensitizing operation applied to the v-cell is a *read* operation: $R \in \{0, 1, ?, -\}$, whereby ? denotes a *random* logic value. A random logic value can occur if the voltage difference between the bit lines (used by the sense amplifier) is very small. A '$-$' in R means that the output data is not applicable; e.g., if $S = w0$, then no data will appear at the memory output, and for that reason R is replaced by a '$-$'. It is interesting to note here that the word *undefined* is used for the state of the cell (for $F =?$), and *random* is used for the read data value (for $R =?$); these words will be used later to give names for the to be introduced FFMs.

Now that the possible values for S, F and R are defined for 1PF1s, it is possible to list all FPs using this notation. Table 3.1 lists all possible combinations of the values, in the $< S/F/R >$ notation, that result in FPs. The remaining combinations of the S, F and R values do not represent a faulty behavior. For example, $< 1w0/0/- >$ corresponds to a correct $w0$ operation after which the cell contains a 0, as expected. The column 'FFM' in the table presents the FFM to which the corresponding FP belongs; such FFMs will be discussed in detail in the next section. Below, the verification of the completeness of the set of 1PF1s shown in Table 3.1 will be proved.

Given that $S \in \{0, 1, 0w0, 1w1, 0w1, 1w0, 0r0, 1r1\}$, $F \in \{0, 1, ?\}$, and $R \in \{0, 1, ?, -\}$, three cases can be distinguished (see Table 3.1):

- $S \in \{0, 1\}$

    if $S = 0$, then $F \in \{1, ?\}$ and $R = -$; this results in FP1 and FP2 of Table 3.1.

    if $S = 1$, then $F \in \{0, ?\}$ and $R = -$; this results in FP3 and FP4.

- $S \in \{0w0, 1w1, 0w1, 1w0\}$

    if $S = 0w0$, then $F \in \{1, ?\}$ and $R = -$; this results in FP5 and FP6.

    if $S = 1w1$, then $F \in \{0, ?\}$ and $R = -$; this results in FP7 and FP8.

    if $S = 0w1$, then $F \in \{0, ?\}$ and $R = -$; this results in FP9 and FP10.

    if $S = 1w0$, then $F \in \{1, ?\}$ and $R = -$; this results in FP11 and FP12.

- $S \in \{0r0, 1r1\}$

    - $S = 0r0$

        if $F = 0$, then $R \in \{1, ?\}$; this results in FP13 and FP14.

        if $F \in \{1, ?\}$, then $R \in \{0, 1, ?\}$; this results in FP15 through FP20.

    - $S = 1r1$

        if $F = 1$, then $R \in \{0, ?\}$; this results in FP21 and FP22.

        if $F \in \{0, ?\}$, then $R \in \{0, 1, ?\}$; this results in FP23 through FP28.

Table 3.1. The complete set of 1PF1 FPs

#	S	F	R	< S/F/R >	FFM	#	S	F	R	< S/F/R >	FFM
1	0	1	–	< 0/1/– >	SF	2	0	?	–	< 0/?/– >	USF
3	1	0	–	< 1/0/– >	SF	4	1	?	–	< 1/?/– >	USF
5	0w0	1	–	< 0w0/1/– >	WDF	6	0w0	?	–	< 0w0/?/– >	UWF
7	1w1	0	–	< 1w1/0/– >	WDF	8	1w1	?	–	< 1w1/?/– >	UWF
9	0w1	0	–	< 0w1/0/– >	TF	10	0w1	?	–	< 0w1/?/– >	UWF
11	1w0	1	–	< 1w0/1/– >	TF	12	1w0	?	–	< 1w0/?/– >	UWF
13	0r0	0	1	< 0r0/0/1 >	IRF	14	0r0	0	?	< 0r0/0/? >	RRF
15	0r0	1	0	< 0r0/1/0 >	DRDF	16	0r0	1	1	< 0r0/1/1 >	RDF
17	0r0	1	?	< 0r0/1/? >	RRDF	18	0r0	?	0	< 0r0/?/0 >	URF
19	0r0	?	1	< 0r0/?/1 >	URF	20	0r0	?	?	< 0r0/?/? >	URF
21	1r1	1	0	< 1r1/1/0 >	IRF	22	1r1	1	?	< 1r1/1/? >	RRF
23	1r1	0	0	< 1r1/0/0 >	RDF	24	1r1	0	1	< 1r1/0/1 >	DRDF
25	1r1	0	?	< 1r1/0/? >	RRDF	26	1r1	?	0	< 1r1/?/0 >	URF
27	1r1	?	1	< 1r1/?/1 >	URF	28	1r1	?	?	< 1r1/?/? >	URF

It is clear from the table that the write operations are capable of sensitizing 8 FPs, and the read operations are capable of sensitizing 16 FPs. Note that in total there are 28 single-cell FPs.

3.3.2 Single-cell functional fault models

The list of 28 possible single-cell FPs will be compiled into a set of FFMs. The FFMs are given names, and each consists of a number of FPs. Selecting which FP should belong to a given generic FFM is rather arbitrary and is mainly determined by historical arguments; see also the columns FFM in Table 3.1, which show the

FFM to which each FP belongs. Table 3.2 summarizes the set of FFMs together with their FPs.

Table 3.2. List of 1PF1 FFMs; $x \in \{0,1\}$

#	FFM	Fault primitives
1	SF	$< 1/0/->, < 0/1/->$
2	TF	$< 0w1/0/->, < 1w0/1/->$
3	WDF	$< 0w0/1/->, < 1w1/0/->$
4	RDF	$< 0r0/1/1 >, < 1r1/0/0 >$
5	DRDF	$< 0r0/1/0 >, < 1r1/0/1 >$
6	RRDF	$< 0r0/1/? >, < 1r1/0/? >$
7	IRF	$< 0r0/0/1 >, < 1r1/1/0 >$
8	RRF	$< 0r0/0/? >, < 1r1/1/? >$
9	USF	$< 1/?/->, < 0/?/->$
10	UWF	$< 0w0/?/->, < 0w1/?/->, < 1w0/?/->, < 1w1/?/->$
11	URF	$< rx/?/0 >, < rx/?/1 >, < rx/?/? >$
12	SAF	$< \forall/0/->, < \forall/1/->$
13	NAF	$\{< 0w1/0/->, < 1w0/1/->, < rx/x/? >\}$
14	DRF	$< 1_T/0/->, < 0_T/1/->, < x_T/?/->$

1. **State Fault (SF)**: A cell is said to have a *state fault* if the logic value of the cell flips before it is accessed, even if no operation is performed on it[1]. This fault is special in the sense that no operation is needed to sensitize it and, therefore, it only depends on the initial stored value in the cell. The SF consists of two FPs: $< 0/1/->$ and $< 1/0/->$.

2. **Transition Fault (TF)**: A cell is said to have a *transition fault* if it fails to undergo a transition ($0 \rightarrow 1$ or $1 \rightarrow 0$) when it is written. This FFM is sensitized by a write operation and depends on both the initial stored logic value and the type of the write operation. The TF consists of two FPs: $< 0w1/0/->$ and $< 1w0/1/->$.

3. **Write Destructive Fault (WDF)**: A cell is said to have a *write destructive fault* if a non-transition write operation ($0w0$ or $1w1$) causes a transition in the cell. The WDF consists of two FPs: $< 0w0/1/->$ and $< 1w1/0/->$.

4. **Read Destructive Fault (RDF)**: A cell is said to have a *read destructive fault* if a read operation performed on the cell changes the data in the cell, and returns an *incorrect* value on the output. The RDF consists of two FPs: $< 0r0/1/1 >$ and $< 1r1/0/0 >$.

---

[1]It should be emphasized here that the state fault should be understood in the static sense. That is, the cell should flip in the short time period after initialization and before accessing the cell.

5. **Deceptive Read Destructive Fault (DRDF)**: A cell is said to have a *deceptive read destructive fault* if a read operation performed on the cell returns the *correct* logic value, while it results in changing the contents of the cell. The DRDF consists of two FPs: $< 0r0/1/0 >$ and $< 1r1/0/1 >$.

6. **Random Read Destructive Fault (RRDF)** : A cell is said to have a *random read destructive fault* if a read operation performed on the cell returns the *random* logic value, while it results in changing the contents of the cell. The RRDF consists of two FPs: $< 0r0/1/? >$ and $< 1r1/0/? >$.

7. **Incorrect Read Fault (IRF)** : A cell is said to have an *incorrect read fault* if a read operation performed on the cell returns the incorrect logic value while keeping the correct stored value in the cell. The IRF consists of two FPs: $< 0r0/0/1 >$ and $< 1r1/1/0 >$.

8. **Random Read Fault (RRF)**: A cell is said to have a *random read fault* if a read operation performed on the cell returns a random data value on the output while the stored value remains as it is. The RRF consists of two FPs: $< 0r0/0/? >$ and $< 1r1/1/? >$.

9. **Undefined State Fault (USF)**: A cell is said to have an *undefined state fault* if the logic value of the cell flips to an undefined state before the cell is accessed, even if no operation is performed on it[1]. This fault is special in the sense that no operation is needed to sensitize it and, therefore, it only depends on the initial stored value in the cell. The USF consists of two FPs: $< 0/?/- >$ and $< 1/?/- >$.

10. **Undefined Write Fault (UWF)**: A cell is said to have an *undefined write fault* if the cell is brought in an undefined state by a write operation. The UWF consists of four FPs: $< 0w0/?/- >$, $< 0w1/?/- >$, $< 1w0/?/- >$, and $< 1w1/?/- >$.

11. **Undefined Read Fault (URF)**: A cell is said to have an *undefined read fault* if the cell is brought in an undefined state by a read operation. The returned data value during this read operation can be correct, wrong, or random. The URF consists of six FPs: $< 0r0/?/0 >$, $< 0r0/?/1 >$, $< 0r0/?/? >$, $< 1r1/?/0 >$, $< 1r1/?/1 >$, and $< 1r1/?/? >$.

12. **Stuck-At Fault (SAF)**: A cell is said to have a *stuck-at fault* if it remains always stuck at a given value for all performed operations. Depending on the value the cell remains stuck at, the SAF has two FPs: $< \forall/0/- >$ and $< \forall/1/- >$. $\forall$ symbolizes all possible sensitizing operations. Therefore, $S = \forall$ can be replaced by any operation that sensitizes the fault. This leads to the following equivalent definitions of the SAF FPs:

- $< \forall/0/- > = \{< 1/0/- >, < 0w1/0/- >, < 1w1/0/- >\}$
- $< \forall/1/- >. = \{< 0/1/- >, < 1w0/1/- >, < 0w0/1/- >\}$

E.g., if the faulty behavior of a cell is said to resemble $< \forall/0/- >$, then the cell exhibits the faulty behavior of each of the three FPs $\{< 1/0/- >, < 0w1/0/- >, < 1w1/0/- >\}$.

13. **No Access Fault (NAF)**: A cell is said to have a *no access fault* if the cell is not accessible; i.e., the state of the cell can not be changed with write operations, and any read operation applied to the cell returns a random data value. The NAF consists of four FPs which occur simultaneously: $\{< 0w1/0/- >, < 1w0/1/- >, < 0r0/0/? >, < 1r1/1/? >\}$. Note that the NAF is a more general form of the *Stuck-Open Fault* [21], which is defined as an inaccessible cell due to an open word line.

14. **Data Retention Fault (DRF)**: A cell is said to have a *data retention fault* if the state of the cell changes after a certain time T, and without accessing the cell. T should be longer than the duration of the precharge cycle in SRAMs, because if the cell flips within the precharge cycle then the sensitized fault would be a state fault. The DRF consists of four FPs: (a) the cell fails to retain the logic value 1 and flips to 0 after certain time T: $< 1_T/0/- >$, (b) the cell fails to retain the logic value 0 and flips to 1 after certain time T: $< 0_T/1/- >$, (c) the cell fails to retain the logic value 1 and becomes in undefined state after certain time T: $< 0_T/?/- >$, and (d) the cell fails to retain the logic value 0 and becomes in undefined state after certain time T: $< 1_T/?/- >$.

Note that the first 11 FFMs defined above contain the 28 possible FPs of Table 3.1. The SAF and the NAF are FFMs which require more than one FP to be present due to the same single defect; they are *composite* FFMs. Note also that the three FFMs *USF*, *UWF* and *URF* bring the cell in an *undefined* state. The letter 'U' in each of these three FFM names is chosen to indicate that effect.

3.3.3   Two-cell fault primitives

Single-port FPs involving two cells (1PF2s) are divided into three types; see Figure 3.2. Before listing the 1PF2s, a precise compact notation for 1PF2s, will be introduced.

$< S_a; S_v/F/R >$ (or $< S_a; S_v/F/R >_{a,v}$): denotes an FP involving two cells; S_a describes the sensitizing operation or state of the aggressor cell (a-cell); while S_v describes the sensitizing operation or state of the victim cell (v-cell). The a-cell (c_a) is the cell sensitizing a fault in an other cell called the v-cell (c_v). The set S_i is defined as: $S_i \in \{0, 1, 0w0, 1w1, 0w1, 1w0, 0r0, 1r1\}$ ($i \in \{a, v\}$), $F \in \{0, 1, ?\}$, and $R \in \{0, 1, ?, -\}$.

Table 3.3 lists all possible combinations of the values, in the $< S_a; S_v/F/R >$ notation, that result in FPs. The column 'FFM' in the table shows the FFM each FP belongs to; such FFMs will be discussed in more detail in the next section. Below, the completeness of the 1PF2s shown in the table will be proved.

Table 3.3. The complete set of 1PF2 FPs; $x \in \{0,1\}$

#	S_a	S_v	F	R	$<S_a,S_v/F/R>$	FFM	#	S_a	S_v	F	R	$<S_a,S_v/F/R>$	FFM
1	x	0	1	–	$<x;0/1/->$	CFst	2	x	0	?	–	$<x;0/?/->$	CFus
3	x	1	0	–	$<x;1/0/->$	CFst	4	x	1	?	–	$<x;1/?/->$	CFus
5	x	0w0	1	–	$<x;0w0/1/->$	CFwd	6	x	0w0	?	–	$<x;0w0/?/->$	CFuw
7	x	1w1	0	–	$<x;1w1/0/->$	CFwd	8	x	1w1	?	–	$<x;1w1/?/->$	CFuw
9	x	0w1	0	–	$<x;0w1/0/->$	CFtr	10	x	0w1	?	–	$<x;0w1/?/->$	CFuw
11	x	1w0	1	–	$<x;1w0/1/->$	CFtr	12	x	1w0	?	–	$<x;1w0/?/->$	CFuw
13	x	0r0	0	1	$<x;0r0/0/1>$	CFir	14	x	0r0	0	?	$<x;0r0/0/?>$	CFrr
15	x	0r0	1	0	$<x;0r0/1/0>$	CFdrd	16	x	0r0	1	1	$<x;0r0/1/1>$	CFrd
17	x	0r0	1	?	$<x;0r0/1/?>$	CFrrd	18	x	0r0	?	0	$<x;0r0/?/0>$	CFur
19	x	0r0	?	1	$<x;0r0/?/1>$	CFur	20	x	0r0	?	?	$<x;0r0/?/?>$	CFur
21	x	1r1	1	0	$<x;1r1/1/0>$	CFir	22	x	1r1	1	?	$<x;1r1/1/?>$	CFrr
23	x	1r1	0	0	$<x;1r1/0/0>$	CFrd	24	x	1r1	0	1	$<x;1r1/0/1>$	CFdrd
25	x	1r1	0	?	$<x;1r1/0/?>$	CFrrd	26	x	1r1	?	0	$<x;1r1/?/0>$	CFur
27	x	1r1	?	1	$<x;1r1/?/1>$	CFur	28	x	1r1	?	?	$<x;1r1/?/?>$	CFur
29	0w0	0	1	–	$<0w0;0/1/->$	CFds	30	0w0	0	?	–	$<0w0;0/?/->$	CFud
31	1w1	0	1	–	$<1w1;0/1/->$	CFds	32	1w1	0	?	–	$<1w1;0/?/->$	CFud
33	0w1	0	1	–	$<0w1;0/1/->$	CFds	34	0w1	0	?	–	$<0w1;0/?/->$	CFud
35	1w0	0	1	–	$<1w0;0/1/->$	CFds	36	1w0	0	?	–	$<1w0;0/?/->$	CFud
37	0r0	0	1	–	$<0r0;0/1/->$	CFds	38	0r0	0	?	–	$<0r0;0/?/->$	CFud
39	1r1	0	1	–	$<1r1;0/1/->$	CFds	40	1r1	0	?	–	$<1r1;0/?/->$	CFud
41	0w0	1	0	–	$<0w0;1/0/->$	CFds	42	0w0	1	?	–	$<0w0;1/?/->$	CFud
43	1w1	1	0	–	$<1w1;1/0/->$	CFds	44	1w1	1	?	–	$<1w1;1/?/->$	CFud
45	0w1	1	0	–	$<0w1;1/0/->$	CFds	46	0w1	1	?	–	$<0w1;1/?/->$	CFud
47	1w0	1	0	–	$<1w0;1/0/->$	CFds	48	1w0	1	?	–	$<1w0;1/?/->$	CFud
49	0r0	1	0	–	$<0r0;1/0/->$	CFds	50	0r0	1	?	–	$<0r0;1/?/->$	CFud
51	1r1	1	0	–	$<1r1;1/0/->$	CFds	52	1r1	1	?	–	$<1r1;1/?/->$	CFud

Given that for the above FP notation, $\#O \leq 1$ (i.e., at the most one operation is allowed to be performed), not all combinations of $S_a \in \{0, 1, 0w0, 1w1, 0w1, 1w0, 0r0, 1r1\}$ and $S_v \in \{0, 1, 0w0, 1w1, 0w1, 1w0, 0r0, 1r1\}$ are possible. Therefore two cases can be distinguished:

- $S_a \in \{0,1\}$: then the notation $< S_a; S_v/F/R >$ represents 28×2 FPs: 28 FPs denoted as $< 0; S_v/F/R >$ and 28 as $< 1; S_v/F/R >$. This notation $< x; S_v/F/R >$ with $x \in \{0,1\}$ is a superset of the notation $< S_v/F/R >$, which represents the 28 1PF1s described in Section 3.3.1, and shown in Table 3.1. Based on the sensitizing operation/state of S_v, we can further classify the 56 FPs into two subclasses:

  – $S_v \in \{0,1\}$: This is the case whereby the *state* of the a-cell sensitizes a fault in the v-cell; i.e., a 1PF2$_s$. This results in eight FPs: FP1 through FP4 of Table 3.3 ($x \in \{0,1\}$).

  – $S_v \in \{0w0, 1w1, 0w1, 1w0, 0r0, 1r1\}$: This is the case whereby the application of a single operation to the v-cell, with the a-cell in a certain state, sensitizes a fault in the v-cell; i.e., a $1PF2_v$. This results in 48 FPs: FP5 through FP28 of Table 3.3.

- $S_a \in \{0w0, 1w1, 0w1, 1w0, 0r0, 1r1\}$: then $S_v \in \{0, 1\}$ and $R = -$. This is the case whereby the application of a single operation to the a-cell sensitizes a fault in the v-cell; i.e., a $1PF2_a$. Depending on the value of S_v, the following cases can be distinguished:

  - if $S_v = 0$, then $F \in \{1, ?\}$. The notation $< S_a; 0/F/- >$ represents $6 \times 2 = 12$ FPs since S_a can be one of the six SOSs, while F can take on one of the two fault effects; see FP29 through FP40 in Table 3.3.
  - if $S_v = 1$, then $F \in \{0, ?\}$. The notation $< S_a; 1/F/- >$ represents $6 \times 2 = 12$ FPs; see FP41 through FP52 in Table 3.3.

Note that the set of possible 1PF2s consists of 80 FPs, of which 8 FPs are $1PF2_s$s, 48 FPs are $1PF2_v$s, and 24 FPs are $1PF2_a$s.

3.3.4 Two-cell functional fault models

The list of 80 possible 1PF2 FPs will be compiled into a set of FFMs. Selecting which FP should belong to a given generic FFM is rather arbitrary and is mainly determined historically; see also the columns 'FFM' in Table 3.3, which show the FFM to which each FP belongs. Table 3.4 summarizes the set of FFMs together with their FPs; each of the FFM will be discussed in detail in the following. Remember that the 1PF2s are divided into three types $1PF2_s$, $1PF2_a$ and $1PF2_v$.

Table 3.4. List of 1PF2 FFMs; $x, y \in \{0, 1\}$

#	FFM	Fault primitives
1	CFst	$< 0; 0/1/- >, < 0; 1/0/- >, < 1; 0/1/- >, < 1; 1/0/- >$
2	CFus	$< 0; 0/?/- >, < 0; 1/?/- >, < 1; 0/?/- >, < 1; 1/?/- >$
3	CFds	$< xwy; 0/1/- >, < xwy; 1/0/- >, < rx; 0/1/- >, < rx; 1/0/- >$
4	CFud	$< xwy; 0/?/- >, < xwy; 1/?/- >, < rx; 0/?/- >, < rx; 1/?/- >$
5	$CFid$	$< 0w1; 0/1/- >, < 0w1; 1/0/- >, < 1w0; 0/1/- >, < 1w0; 1/0/- >$
6	$CFin$	$\{< 0w1; 0/1/- >, < 0w1; 1/0/- >\}, \{< 1w0; 0/1/- >, < 1w0; 1/0/- >\}$
7	CFtr	$< 0; 0w1/0/- >, < 1; 0w1/0/- >, < 0; 1w0/1/- >, < 1; 1w0/1/- >$
8	CFwd	$< 0; 0w0/- >, < 1; 0w0/1/- >, < 0; 1w1/0/- >, < 1; 1w1/0/- >$
9	CFrd	$< 0; 0r0/1/1 >, < 1; 0r0/1/1 >, < 0; 1r1/0/0 >, < 1; 1r1/0/0 >$
10	CFdrd	$< 0; 0r0/1/0 >, < 1; 0r0/1/0 >, < 0; 1r1/0/1 >, < 1; 1r1/0/1 >$
11	CFrrd	$< 0; 0r0/1/? >, < 1; 0r0/1/? >, < 0; 1r1/0/? >, < 1; 1r1/0/? >$
12	CFir	$< 0; 0r0/0/1 >, < 1; 0r0/0/1 >, < 0; 1r1/1/0 >, < 1; 1r1/1/0 >$
13	CFrr	$< 0; 0r0/0/? >, < 1; 0r0/0/? >, < 0; 1r1/1/? >, < 1; 1r1/1/? >$
14	CFuw	$< x; 0w0/?/- >, < x; 0w1/?/- >, < x; 1w0/?/- >, < x; 1w1/?/- >,$
15	CFur	$< x; 0r0/?/0 >, < x; 0r0/?/1 >, < x; 0r0/?/? >,$ $< x; 1r1/?/1 >, < x; 1r1/?/0 >, < x; 1r1/?/? >$

The 1PF2$_s$ FFMs

This type has the property that the *state* of the a-cell, rather than an operation applied to the a-cell, sensitizes a fault in the v-cell; it consists of two FFMs; see Table 3.4:

1. **State coupling fault (CFst)**: Two cells are said to have a *state coupling fault* if the v-cell is forced into a given logic state only if the a-cell is in a given state, without performing any operation on the v-cell or on the a-cell. This fault is special in the sense that no operation is needed to sensitize it and, therefore, it only depends on the initial stored values in the cells. The CFst consists of four FPs: $< 0; 0/1/- >$, $< 0; 1/0/- >$, $< 1; 0/1/- >$, and $< 1; 1/0/- >$.

2. **Undefined State coupling fault (CFus)**: Two cells are said to have an *undefined state coupling fault* if the v-cell is forced into an undefined logic state only if the a-cell is in a given state, without performing any operation on the v-cell. The CFus consists of four FPs: $< 0; 0/?/- >$, $< 0; 1/?/- >$, $< 1; 0/?/- >$, and $< 1; 1/?/- >$.

The 1PF2$_a$ FFMs

This type has the property that the application of a single-port operation to the a-cell sensitizes a fault in the v-cell; it consists of the following FFMs; see Table 3.4:

1. **Disturb coupling fault (CFds)**: Two cells are said to have a *disturb coupling fault* if an operation (write or read) performed on the a-cell causes the v-cell to flip. Here, any operation performed on the a-cell is accepted as a sensitizing operation for the fault, be it a read, a transition write or a non-transition write operation. The CFds consists of 12 FPs: $< 0w0; 0/1/- >$, $< 1w1; 0/1/- >$, $< 0w1; 0/1/- >$, $< 1w0; 0/1/- >$, $< 0r0; 0/1/- >$, $< 1r1; 0/1/- >$, $< 0w0; 1/0/- >$, $< 1w1; 1/0/- >$, $< 0w1; 1/0/- >$, $< 1w0; 1/0/- >$, $< 0r0; 1/0/- >$, and $< 1r1; 1/0/- >$.

2. **Undefined Disturb coupling fault (CFud)**: Two cells are said to have an *undefined disturb coupling fault* if an operation performed on the a-cell forces the v-cell into an undefined logic state. Here, any operation performed on the a-cell is accepted as a sensitizing operation for the fault, be it a read, a transition write or a non-transition write operation. The CFud consists of 12 FPs: $< 0w0; 0/?/- >$, $< 1w1; 0/?/- >$, $< 0w1; 0/?/- >$, $< 1w0; 0/?/- >$, $< 0r0; 0/?/- >$, $< 1r1; 0/?/- >$, $< 0w0; 1/?/- >$, $< 1w1; 1/?/- >$, $< 0w1; 1/?/- >$, $< 1w0; 1/?/- >$, $< 0r0; 1/?/- >$, and $< 1r1; 1/?/- >$.

3. **Idempotent coupling fault (CFid)**: Two cells are said to have an *idempotent coupling fault* if a transition write operation ($0w1$ and $1w0$) on the a-cell causes the v-cell to flip. This fault is sensitized by a *transition write* operation

performed on the a-cell. The CFid consists of four FPs: $< 0w1; 0/1/- >$, $< 0w1; 1/0/- >$, $< 1w0; 0/1/- >$, and $< 1w0; 1/0/- >$.

4. **Inversion coupling fault (CFin)**: Two cells are said to have an *inversion coupling fault* if the logic value of the v-cell is inverted in case a *transition write* operation is performed on the a-cell. The CFin consists of two pairs of FPs; the two FPs of each pair have to be present simultaneously: $\{< 0w1; 0/1/- >, < 0w1; 1/0/- >\}$ and $\{< 1w0; 0/1/- >, < 1w0; 1/0/- >\}$. The $\{< 0w1; 0/1/- >, < 0w1; 1/0/- >\}$ describes the fact that the transition operation '$0w1$' applied to the a-cell results in setting the state of the v-cell to 1 if it contains a 0, and to 0 if it contains a 1.

The 1PF2$_v$ FFMs

This type has the property that the application of a single-port operation to the v-cell (with the a-cell in certain state) sensitizes a fault in the v-cell. It consists of the following FFMs; see Table 3.4:

1. **Transition coupling fault (CFtr)**: Two cells are said to have a *transition coupling fault* if a given logic value in the aggressor results in a failing transition write operation performed on the victim. This fault is sensitized by first setting the a-cell in a given state, and thereafter applying a write operation on the v-cell. The CFtr consists of four FPs: $< 0; 0w1/0/- >$, $< 1; 0w1/0/- >$, $< 0; 1w0/1/- >$, and $< 1; 1w0/1/- >$.

2. **Write Destructive coupling fault (CFwd)**: Two cells are said to have a *write destructive coupling fault* if a non-transition write operation performed on the v-cell results in a transition when the a-cell is in a given logic state. The CFwd consists of four FPs: $< 0; 0w0/1/- >$, $< 1; 0w0/1/- >$, $< 0; 1w1/0/- >$, and $< 1; 1w1/0/- >$.

3. **Read Destructive coupling fault (CFrd)**: Two cells are said to have a *read destructive coupling fault* when a read operation performed on the v-cell changes the data in the v-cell and returns an *incorrect* value on the output, if the a-cell is in a given state. The CFrd consists of four FPs: $< 0; 0r0/1/1 >$, $< 1; 0r0/1/1 >$, $< 0; 1r1/0/0 >$, and $< 1; 1r1/0/0 >$.

4. **Deceptive Read Destructive coupling fault (CFdrd)**: Two cells are said to have a *deceptive read destructive coupling fault* when a read operation performed on the v-cell changes the data in the v-cell and returns a *correct* value on the output, if the a-cell is in a given state. The CFdrd consists of four FPs: $< 0; 0r0/1/0 >$, $< 1; 0r0/1/0 >$, $< 0; 1r1/0/1 >$, and $< 1; 1r1/0/1 >$.

5. **Random Read Destructive coupling fault (CFrrd)**: Two cells are said to have a *random read destructive coupling fault* when a read operation performed

on the v-cell changes the data in the cell and returns a *random* value on the output, if the a-cell is in a given state. The CFrrd consists of four FPs: $< 0; 0r0/1/? >$, $< 1; 0r0/1/? >$, $< 0; 1r1/0/? >$, and $< 1; 1r1/0/? >$.

6. **Incorrect Read coupling fault (CFir)**: Two cells are said to have an *incorrect read coupling fault* if a read operation performed on the v-cell returns the incorrect logic value when the a-cell is in a given state. Note here that the state of the v-cell is not changed. The CFir consists of four FPs: $< 0; 0r0/0/1 >$, $< 1; 0r0/0/1 >$, $< 0; 1r1/1/0 >$, and $< 1; 1r1/1/0 >$.

7. **Random Read coupling fault (CFrr)**: Two cells are said to have a *random read coupling fault* if a read operation performed on the v-cell changes the data in the v-cell and returns a *correct* value on the output, when the a-cell is in a given state. Note that the state of the v-cell is not impacted; it remains in its correct value. The CFrr consists of four FPs: $< 0; 0r0/0/? >$, $< 1; 0r0/0/? >$, $< 0; 1r1/1/? >$, and $< 1; 1r1/1/? >$.

8. **Undefined Write coupling fault (CFuw)**: Two cells are said to have an *undefined write coupling fault* if the v-cell is brought in an undefined state by a write operation performed on the v-cell, when the a-cell is in a given state. The CFuw consists of eight FPs: $< x; 0w0/?/- >$, $< x; 0w1/?/- >$, $< x; 1w0/?/- >$, and $< x; 1w1/?/- >$; $x \in \{0, 1\}$.

9. **Undefined Read coupling fault (CFur)**: Two cells are said to have an *undefined read coupling fault* if the v-cell is brought in an undefined state by a read operation performed on the v-cell, when the a-cell is in a given state. The returned data value during this read operation can be correct, wrong, or random. The CFur consists of twelve FPs: $< x; 0r0/?/0 >$, $< x; 0r0/?/1 >$, $< x; 0r0/?/? >$, $< x; 1r1/?/1 >$, $< x; 1r1/?/0 >$, and $< x; 1r1/?/? >$; $x \in \{0, 1\}$.

There is a need to select a collection of the FFMs presented above that would cover all FPs listed in Table 3.3. An analysis of the defined FFMs shows that all introduced FFMs are necessary except the CFid and CFin. These two FFMs have been introduced for historical reasons.

3.4   Two-port fault primitives

Two-port faults (2PFs) can not occur in single-port memories; they require the use of two ports simultaneously in order to be sensitized. They are considered only for memories with two or more ports; i.e., $p \geq 2$. They are divided into single-cell FPs and multi-cell FPs. Single-cell FPs involve a single cell; while multi-cell FPs involve more than one cell. We will restrict our analysis to multi-cell FPs involving at most three cells (i.e., $\#C \leq 3$). Figure 3.3 shows the three classes of the 2PFs, whereby $\#C \leq 3$:

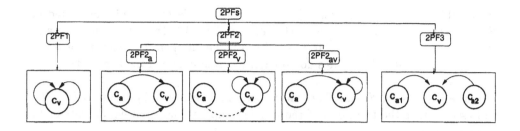

Figure 3.3. Classification of 2PFs

1. *The 2PFs involving a single cell (2PF1s):* They are based on a combination of two single-port operations applied simultaneously via two ports to a single cell; the cell accessed is the same cell as where the fault appears.

2. *The 2PFs involving two cells (2PF2s):* Depending to which cells the two simultaneous operations are applied (to the a-cell, and/or to the v-cell), the 2PF2s can be further divided into three types:

    1. The $2PF2_a$: this fault is sensitized in the v-cell c_v by applying two simultaneous operations to the same a-cell c_a.
    2. The $2PF2_v$: this fault is sensitized in the v-cell by applying two simultaneous operations to the same v-cell (solid arrows in Figure 3.3), while the a-cell has to be in a certain state (dashed arrow in the figure).
    3. The $2PF2_{av}$: this fault is sensitized in the v-cell by applying two simultaneous operations: one to the a-cell and one to the v-cell.

3. *The 2PFs involving three cells (2PF3s):* The fault is sensitized in the v-cell when two simultaneous operations are applied via two different ports to two different a-cells.

In the remainder of this section, the domain of all possible FPs for 2PF1s, 2PF2s as well as for 2PF3s will be presented.

3.4.1   Single-cell fault primitives

Before listing the possible 2PF1s, the to be used FP notation will be introduced.

$< S1 : S2/F/R >_v$ denotes a two-port FP involving a single cell (v-cell). This FP requires the use of the two ports *simultaneously*. $S1$ and $S2$ describe the sensitizing operations of the cell; ":" denotes the fact that $S1$ and $S2$ are applied *simultaneously* through the two ports. F describes the value of the v-cell. Note that the sensitizing operations are applied to the same cell as where the fault appears. R is the read result of $S1$ and $S2$ in the case they are both the *same* read operation (i.e., '0r0 : 0r0' or '1r1 : 1r1'). In the case one of the read operations returns the expected value, while the other returns a wrong value, then the latter value is considered in R. The sets are given as: $Si \in \{0w0, 1w1, 0w1, 1w0, 0r0, 1r1\}$, $F \in \{0, 1, ?\}$, and $R \in \{0, 1, ?, -\}$; $i \in \{1, 2\}$. Note that Si has to be an operation, and not a state; because FPs sensitized by states do no require acting upon via the two ports.

Since the sets Si, F, and R are know, one can determine all possible 2PF1s. We will assume that the two ports of the memory support the following operations to the *same location*:

- Two simultaneous reads; i.e., $(S1 : S2) \in \{(0r0 : 0r0), (1r1 : 1r1)\}$.

- Simultaneous read and write. However, in that case the read data will be discarded; i.e, the write operation has a high priority. That is $(S1 : S2) \in \{(0r0 : 0w0), (0r0 : 0w1), (1r1 : 1w1), (1r1 : 1w0)\}$. Note that $(S1 : S2)$ can also one of the following $\{(0w0 : 0r0), (0w1 : 0r0), (1w1 : 1r1), (1w0 : 1r1)\}$; however, we will only restrict ourself to the first set of operations due to the following assumptions:

  - If a fault is caused by a defect *within a cell*; i.e., not port related, then, e.g., the $(0r0 : 0w1)$ will sensitize the same fault as $(0w1 : 0r0)$.

  - If a fault is caused by a defect *between two ports*, or *between two cells, then is assumed that the defect (e.g., extra undesired metal) is symmetrical with respect with the two ports, respectively, the two cells.* Therefore, the fault that will be sensitized by, e.g., $(0r0 : 0w1)$ is similar to that sensitized by $(0w1 : 0r0)$.

The validity of these two assumptions has been verified by using Inductive Fault Analysis and circuit simulation.

- *No* simultaneous write operations are supported.

Table 3.5 lists all possible combinations of the values, in the $< S1 : S2/F/R >_v$ notation, that result in FPs. The column 'FFM' gives the FFM to which each FP belongs; these FFMs will be presented in more detail in the next section. However, first the verification of the completeness of 2PF1s shown in the table will be proven.

- $(S1 : S2) \in \{(0r0 : 0r0), (1r1 : 1r1)\}$:

  - '$S1 : S2 = 0r0 : 0r0$':
    If $F = 0$, then $R \in \{1, ?\}$; this results in FP1 and FP2; see Table 3.6.
    If $F \in \{1, ?\}$, then $R \in \{0, 1, ?\}$; this results in FP3 through FP8.

  - '$S1 : S2 = 1r1 : 1r1$':
    If $F = 1$, then $R \in \{0, ?\}$; this results in FP9 and FP10.
    If $F \in \{0, ?\}$, then $R \in \{0, 1, ?\}$; this results in FP11 through FP16.

- $(S1:S2) \in \{(0r0:0w0),(0r0:0w1),(1r1:1w1),(1r1:1w0)\}$:

  - If '$S1:S2 = 0r0:0w0$', then $F \in \{1,?\}$ and $R = -$; this results in FP17 and FP18.

  - If '$S1:S2 = 0r0:0w1$', then $F \in \{0,?\}$ and $R = -$; this results in FP19 and FP20.

  - If '$S1:S2 = 1r1:1w1$', then $F \in \{0,?\}$ and $R = -$; this results in FP21 and FP22.

  - If '$S1:S2 = 1r1:1w0$', then $F \in \{1,?\}$ and $R = -$; this results in FP23 and FP24.

Table 3.5. The complete set of 2PF1 FPs

#	$S1:S2$	F	R	$<S1:S2/F/R>$	FFM
1	$0r0:0r0$	0	1	$<0r0:0r0/0/1>$	wIRF&wIRF
2	$0r0:0r0$	0	?	$<0r0:0r0/0/?>$	wRRF&wRRF
3	$0r0:0r0$	1	0	$<0r0:0r0/1/0>$	wDRDF&wDRDF
4	$0r0:0r0$	1	1	$<0r0:0r0/1/1>$	wRDF&wRDF
5	$0r0:0r0$	1	?	$<0r0:0r0/1/?>$	wRRDF&wRRDF
6	$0r0:0r0$	?	0	$<0r0:0r0/?/0>$	wURF&wURF
7	$0r0:0r0$	?	1	$<0r0:0r0/?/1>$	wURF&wURF
8	$0r0:0r0$	?	?	$<0r0:0r0/?/?>$	wURF&wURF
9	$1r1:1r1$	1	0	$<1r1:1r1/1/0>$	wIRF&wIRF
10	$1r1:1r1$	1	?	$<1r1:1r1/1/?>$	wRRF&wRRF
11	$1r1:1r1$	0	0	$<1r1:1r1/0/1>$	wDRDF&wDRDF
12	$1r1:1r1$	0	1	$<1r1:1r1/0/0>$	wRDF&wRDF
13	$1r1:1r1$	0	?	$<1r1:1r1/0/?>$	wRRDF&wRRDF
14	$1r1:1r1$	?	0	$<1r1:1r1/?/1>$	wURF&wURF
15	$1r1:1r1$	?	1	$<1r1:1r1/?/0>$	wURF&wURF
16	$1r1:1r1$	?	?	$<1r1:1r1/?/?>$	wURF&wURF
17	$0r0:0w0$	1	–	$<0r0:0w0/1/->$	wRDF&wWDF
18	$0r0:0w0$	?	–	$<0r0:0w0/?/->$	wURF&wUWF
19	$0r0:0w1$	0	–	$<0r0:0w1/0/->$	wRDF&wTF
20	$0r0:0w1$	?	–	$<0r0:0w1/?/->$	wURF&wUWF
21	$1r1:1w1$	0	–	$<1r1:1w1/0/->$	wRDF&wWDF
22	$1r1:1w1$	?	–	$<1r1:1w1/?/->$	wURF&wUWF
23	$1r1:1w0$	1	–	$<1r1:1w0/1/->$	wRDF&wTF
24	$1r1:1w0$	?	–	$<1r1:1w0/?/->$	wURF&wUWF

3.4.2 Single-cell functional fault models

The list of 24 possible single-cell FPs shown in Table 3.5 will be compiled into a set of FFMs. Before introducing the FFM, the following terminology will be (re)introduced [8, 27, 63, 92]:

- *Strong fault:* This is a memory fault that can be **fully** *sensitized* by an operation; e.g., a single-port (SP) write or read operation fails, two simultaneous

read operations fail, etc. That means that the state of the v-cell is incorrectly
changed, can not be changed, or that the output(s) of the memory return(s)
an incorrect result(s).

- *Weak fault:* This is a fault which is **partially** sensitized by an operation; e.g.,
  due to a defect that creates a small disturbance of the voltage of the true node
  of the cell. However, a fault can be *fully sensitized* (i.e., becomes strong) when
  two (or more) weak faults are sensitized simultaneously, since their fault effects
  can be additive. This may occur when a p-port (pP) operation is applied. Note
  that in the presence of a weak fault, all SP (read and write) operations pass
  correctly, and that the pP operations may pass correctly. The latter will be the
  case if the fault effects of the weak faults are not sufficient to fully sensitize a
  fault.

The terminology of weak and strong faults is used in representing multi-port
FFMs as follows:

- F denotes a *strong fault* F, while wF denotes the *weak fault* F. For example,
  RDF denotes a strong Read Destructive Fault, while $wRDF$ denotes a weak
  Read Destructive Fault.

- $< fault_1 > \& < fault_2 > ...\& < fault_p >$: denotes a pPF consisting of p weak
  faults; '&' denotes the fact that the p faults *in parallel* (i.e., simultaneously)
  form the pPF. E.g., the wRDF&wRDF denote a 2PF based on two weak RDFs.

Table 3.6 shows the two-port, single-cell FFMs together with their FPs; see also
column 'FFM' in Table 3.5. They cover all 24 FPs of Table 3.5. The introduced
FFMs are explained below.

Table 3.6. List of 2PF1 FFMs; $x \in \{0,1\}$

#	FFM	Fault primitives
1	wRDF&wRDF	$< 0r0 : 0r0/1/1 >, < 1r1 : 1r1/0/0 >$
2	wDRDF&wDRDF	$< 0r0 : 0r0/1/0 >, < 1r1 : 1r1/0/1 >$
3	wRRDF&wRRDF	$< 0r0 : 0r0/1/? >, < 1r1 : 1r1/0/? >$
4	wIRF&wIRF	$< 0r0 : 0r0/0/1 >, < 1r1 : 1r1/1/0 >$
5	wRRF&wRRF	$< 0r0 : 0r0/0/? >, < 1r1 : 1r1/1/? >$
6	wURF&wURF	$< rx : rx/?/0 >, < rx : rx/?/1 >, < rx : rx/?/? >$
7	wRDF&wTF	$< 0r0 : 0w1/0/- >, < 1r1: 1w0/1/- >$
8	wRRF&wWDF	$< 0r0 : 0w0/1/- >, < 1r1 : 1w1/0/- >$
9	wURF&wUWF	$< 0r0 : 0w0/?/- >, < 0r0 : 0w1/?/- >, < 1r1 : 1w1/?/- >, < 1r1 : 1w1/?/- >$

1. wRDF&wRDF: Applying two simultaneous read operations to the v-cell causes
   the v-cell to flip and the sense amplifiers return *incorrect* values.

2. wDRDF&wDRDF: Applying two simultaneous read operations to the v-cell causes the v-cell to flip, while the sense amplifiers return *correct values.*

3. wRRDF&wRRDF: Applying two simultaneous read operations to the v-cell causes the v-cell to flip, while the sense amplifiers return *random* values.

4. wIRF&wIRF: Applying two simultaneous read operations to the v-cell return *incorrect* logic values, while the cell maintains its correct stored value.

5. wRRF&wRRF: Applying two simultaneous read operations to the v-cell return *random* logic values, while the v-cell maintains its correct stored value.

6. wURF&wURF: Applying two simultaneous read operations to the v-cell bring the v-cell in an *undefined* state. The data values returned by the simultaneous read operations can be correct, wrong, or random.

7. wRDF&wTF: The v-cell fails to undergo a write transition if a read operation is applied to the v-cell simultaneously.

8. wRRF&wWDF: Applying simultaneously a read and a non-transition write operation to the v-cell cause the same cell to flip.

9. wURF&wUWF: Applying simultaneously a read and a write to the v-cell causes an *undefined* state in v-cell.

3.4.3   Two-cell fault primitives

The two-port, two-cell FPs (2PF2s) have been divided, depending to which cells the two simultaneous operations are applied (to the a-cell, and/or to the v-cell), into three types (see Figure 3.3). Below each type will be discussed separately.

The 2PF2$_a$ FPs

A 2PF2$_a$ FP is sensitized by applying two simultaneous operations to the same a-cell; see Figure 3.3. To denote the 2PF2$_a$, the following FP notation will be used:
$< S1_a : S2_a; S_v/F/R >_{a,v}$. It denotes an FP whereby both sensitizing operations, $S1_a$ and $S2_a$, are applied simultaneously to the a-cell. S_v denotes the state of the v-cell; i.e., $S_v \in \{0, 1\}$. F denotes the value of the faulty cell c_v. Note that R will be replaced with '$-$' since S_v can not be a read operation.

Table 3.7 lists the 24 possible 2PF2$_a$ FPs. Note that $S1_a$ and $S2_a$ can only be sensitizing *operations*, not states. This consists of six simultaneous pairs of operations to the same cell, assumed to be supported by the two ports of the memory; i.e., $(S1_a, S2_a) \in \{(0r0 : 0r0), (1r1 : 1r1), (0r0 : 0w0), (0r0 : 0w1), (1r1 : 1w1), (1r1 : 1w0)\}$. Here, the same assumption will be made as that for 2PF1s (see Section

3.4.1); e.g., the '$(1r1 : 1w0)$' and '$(1w0 : 1r1)$' cause similar faults. In addition, $S_v \in \{0, 1\}$:

- If $S_v = 0$, then $F \in \{1, ?\}$ and $R = -$; the notation $< S1_a : S2_a; S_v/F/R >_{a,v}$
  results in $6 \times 2 = 12$ FPs, since '$S1_a : S2_a$' can take on one of the six possible
  simultaneous pairs of operations and F can take on one of the two possible
  fault effects (FP1 through FP12 in Table 3.7).

- If $S_v = 1$, then $F \in \{0, ?\}$ and $R = -$; the notation $< S1_a : S2_a; S_v/F/R >_{a,v}$
  results in 12 FPs (FP13 through FP24 in Table 3.7).

Table 3.7. The compete set of $2PF2_a$ FPs

#	$S1_a : S2_a$	S_v	F	R	$< S1_a : S2_a; S_v/F/R >$	FFM
1	$0r0 : 0r0$	0	1	–	$< 0r0 : 0r0; 0/1/- >$	wCFds&wCFds
2	$0r0 : 0r0$	0	?	–	$< 0r0 : 0r0; 0/?/- >$	wCFud&wCFud
3	$1r1 : 1r1$	0	1	–	$< 1r1 : 1r1; 0/1/- >$	wCFds&wCFds
4	$1r1 : 1r1$	0	?	–	$< 1r1 : 1r1; 0/?/- >$	wCFud&wCFud
5	$0r0 : 0w0$	0	1	–	$< 0r0 : 0w0; 0/1/- >$	wCFds&wCFds
6	$0r0 : 0w0$	0	?	–	$< 0r0 : 0w0; 0/?/- >$	wCFud&wCFud
7	$0r0 : 0w1$	0	1	–	$< 0r0 : 0w1; 0/1/- >$	wCFds&wCFds
8	$0r0 : 0w1$	0	?	–	$< 0r0 : 0w1; 0/?/- >$	wCFud&wCFud
9	$1r1 : 1w1$	0	1	–	$< 0r0 : 1w1; 0/1/- >$	wCFds&wCFds
10	$1r1 : 1w1$	0	?	–	$< 0r0 : 1w1; 0/?/- >$	wCFud&wCFud
11	$1r1 : 1w0$	0	1	–	$< 1r1 : 1w0; 0/1/- >$	wCFds&wCFds
12	$1r1 : 1w0$	0	?	–	$< 1r1 : 1w0; 0/?/- >$	wCFud&wCFud
13	$0r0 : 0r0$	1	0	–	$< 0r0 : 0r0; 1/0/- >$	wCFds&wCFds
14	$0r0 : 0r0$	1	?	–	$< 0r0 : 0r0; 1/?/- >$	wCFud&wCFud
15	$1r1 : 1r1$	1	0	–	$< 1r1 : 1r1; 1/0/- >$	wCFds&wCFds
16	$1r1 : 1r1$	1	?	–	$< 1r1 : 1r1; 1/?/- >$	wCFud&wCFud
17	$0r0 : 0w0$	1	0	–	$< 0r0 : 0w0; 1/0/- >$	wCFds&wCFds
18	$0r0 : 0w0$	1	?	–	$< 0r0 : 0w0; 1/?/- >$	wCFud&wCFud
19	$0r0 : 0w1$	1	0	–	$< 0r0 : 0w1; 1/0/- >$	wCFds&wCFds
20	$0r0 : 0w1$	1	?	–	$< 0r0 : 0w1; 1/?/- >$	wCFud&wCFud
21	$1r1 : 1w1$	1	0	–	$< 0r0 : 1w1; 1/0/- >$	wCFds&wCFds
22	$1r1 : 1w1$	1	?	–	$< 0r0 : 1w1; 1/?/- >$	wCFud&wCFud
23	$1r1 : 1w0$	1	0	–	$< 1r1 : 1w0; 1/0/- >$	wCFds&wCFds
24	$1r1 : 1w0$	1	?	–	$< 1r1 : 1w0; 1/?/- >$	wCFud&wCFud

The $2PF2_v$ FPs

A $2PF2_v$ FP is sensitized in the v-cell by applying two simultaneous operations to
the same v-cell, while the a-cell has to be in certain state; see Figure 3.3. To denote
the $2PF2_v$, the following FP notation will be used: $< S_a; S1_v : S2_v/F/R >_{a,v}$. It
denotes an FP whereby both sensitizing operations, $S1_v$ and $S2_v$, are applied simultaneously to the v-cell. S_a describes the state of the a-cell; i.e., $S_a \in \{0, 1\}$.

Table 3.8 lists all possible 2PF2$_v$ faults; they consist of 48 FPs. Note that $S1_v$ and $S2_v$ can only be one of the six possible pairs of two simultaneous sensitizing *operations*, not states. Since $S_a \in \{0, 1\}$, the notation $< S_a \; ; \; S1_v : S2_v/F/R >$ represents 24×2 FPs: 24 FPs denoted as $< 0 \; ; \; S1_v : S2_v/F/R >$, and 24 as $< 1 \; ; \; S1_v : S2_v/F/R >$. Note that in that representation, $< S1_v : S2_v/F/R >$ represents the 24 2PF1s described in Section 3.4.1, and given in Table 3.5.

Table 3.8. The complete set of 2PF2$_v$ FPs; $x \in \{0, 1\}$

#	S_a	$S1_v : S2_v$	F	R	$< S_a; S1_v : S2_v/F/R >$	FFM
1	x	$0r0 : 0r0$	0	1	$< x; 0r0 : 0r0/0/1 >$	wCFir&wIRF
2	x	$0r0 : 0r0$	0	?	$< x; 0r0 : 0r0/0/? >$	wCFrr&wRRF
3	x	$0r0 : 0r0$	1	0	$< x; 0r0 : 0r0/1/0 >$	wCFdrd&wDRDF
4	x	$0r0 : 0r0$	1	1	$< x; 0r0 : 0r0/1/1 >$	wCFrd&wRDF
5	x	$0r0 : 0r0$	1	?	$< x; 0r0 : 0r0/1/? >$	wCFrrd&wRRDF
6	x	$0r0 : 0r0$	?	0	$< x; 0r0 : 0r0/?/0 >$	wCFur&wURF
7	x	$0r0 : 0r0$	?	1	$< x; 0r0 : 0r0/?/1 >$	wCFur&wURF
8	x	$0r0 : 0r0$	?	?	$< x; 0r0 : 0r0/?/? >$	wCFur&wURF
9	x	$1r1 : 1r1$	1	0	$< x; 1r1 : 1r1/1/0 >$	wCFir&wIRF
10	x	$1r1 : 1r1$	1	?	$< x; 1r1 : 1r1/1/? >$	wCFrr&wRRF
11	x	$1r1 : 1r1$	0	1	$< x; 1r1 : 1r1/0/1 >$	wCFdrd&wDRDF
12	x	$1r1 : 1r1$	0	1	$< x; 1r1 : 1r1/0/0 >$	wCFrd&wRDF
13	x	$1r1 : 1r1$	0	?	$< x; 1r1 : 1r1/0/? >$	wCFrrd&wRRDF
14	x	$1r1 : 1r1$	?	1	$< x; 1r1 : 1r1/?/1 >$	wCFur&wURF
15	x	$1r1 : 1r1$	?	0	$< x; 1r1 : 1r1/?/0 >$	wCFur&wURF
16	x	$1r1 : 1r1$	?	?	$< x; 1r1 : 1r1/?/? >$	wCFur&wURF
17	x	$0r0 : 0w0$	1	–	$< x; 0r0 : 0w0/1/– >$	wCFrd&wWDF
18	x	$0r0 : 0w0$	?	–	$< x; 0r0 : 0w0/?/– >$	wCFur&wUWF
19	x	$0r0 : 0w1$	0	–	$< x; 0r0 : 0w1/0/– >$	wCFrd&wTF
20	x	$0r0 : 0w1$	?	–	$< x; 0r0 : 0w1/?/– >$	wCFur&wUWF
21	x	$1r1 : 1w1$	0	–	$< x; 1r1 : 1w1/0/– >$	wCFrd&wWDF
22	x	$1r1 : 1w1$	?	–	$< x; 1r1 : 1w1/?/– >$	wCFur&wUWF
23	x	$1r1 : 1w0$	1	–	$< x; 1r1 : 1w0/1/– >$	wCFrd&wTF
24	x	$1r1 : 1w0$	?	–	$< x; 1r1 : 1w0/?/– >$	wCFur&wUWF

The 2PF2$_{av}$ FPs

A 2PF2$_{av}$ FP is sensitized in the v-cell by applying two simultaneous operations: one to the a-cell and one to the v-cell; see Figure 3.3. To denote the 2PF2$_{av}$, the following FP notation will be used:

$< S_a : S_v/F/R >_{a,v}$. It denotes an FP whereby the sensitizing operation S_a is applied to the a-cell simultaneously with the sensitizing operation S_v, which is applied to the v-cell.

S_a and S_v are both sensitizing operations; i.e., $S_a, S_v \in \{0r0, 1r1, 0w0, 1w1, 0w1, 1w0\}$. In the notation $< S_a : S_v/F/R >_{a,v}$, the $< S_v/F/R >$ represents the same FPs as

1PF1s discussed in Section 3.3.1, except for the FPs whereby $S_v \in \{0,1\}$; they consist of 24 FPs (FP5 through FP28 in Table 3.1). Since S_a can be one of the six operations (i.e., $S_a \in \{0r0, 1r1, 0w0, 1w1, 0w1, 1w0\}$), the notation $< S_a : S_v/F/R >_{a,v}$ represents $6 \times 24 = 144$ FPs, listed in Table 3.9.

Table 3.9. The complete set of 2PF$_{av}$ FPs; $op \in \{0r0, 1r1, 0w0, 1w1, 0w1, 1w0\}$

#	S_a	S_v	F	R	$< S_a : S_v/F/R >$	FFM
1	op	$0w0$	1	–	$< op : 0w0/1/- >$	wCFds&wWDF
2	op	$0w0$	?	–	$< op : 0w0/?/- >$	wCFds&wUWF
3	op	$1w1$	0	–	$< op : 1w1/0/- >$	wCFds&wWDF
4	op	$1w1$	?	–	$< op : 0w0/?/- >$	wCFds&wUWF
5	op	$0w1$	0	–	$< op : 0w1/0/- >$	wCFds&wTF
6	op	$0w1$	?	–	$< op : 0w1/?/- >$	wCFds&wUWF
7	op	$1w0$	1	–	$< op : 1w0/1/- >$	wCFds&wTF
8	op	$1w0$	?	–	$< op : 1w1/?/- >$	wCFds&wUWF
9	op	$0r0$	0	1	$< op : 0r0/0/1 >$	wCFds&wIRF
10	op	$0r0$	0	?	$< op : 0r0/0/? >$	wCFds&wRRF
11	op	$0r0$	1	0	$< op : 0r0/1/0 >$	wCFds&wDRDF
12	op	$0r0$	1	1	$< op : 0r0/1/1 >$	wCFds&wRDF
13	op	$0r0$	1	?	$< op : 0r0/1/? >$	wCFds&wRRDF
14	op	$0r0$	?	0	$< op : 0r0/?/0 >$	wCFds&wURF
15	op	$0r0$	?	1	$< op : 0r0/?/1 >$	wCFds&wURF
16	op	$0r0$	?	?	$< op : 0r0/?/? >$	wCFds&wURF
17	op	$1r1$	1	0	$< op : 1r1/1/0 >$	wCFds&wIRF
18	op	$1r1$	1	?	$< op : 1r1/1/? >$	wCFds&wRRF
19	op	$1r1$	0	0	$< op : 1r1/0/0 >$	wCFds&wRDF
20	op	$1r1$	0	1	$< op : 1r1/0/1 >$	wCFds&wDRDF
21	op	$1r1$	0	?	$< op : 1r1/0/? >$	wCFds&wRRDF
22	op	$1r1$	?	0	$< op : 1r1/?/0 >$	wCFds&wURF
23	op	$1r1$	?	1	$< op : 1r1/?/1 >$	wCFds&wURF
24	op	$1r1$	?	?	$< op : 1r1/?/? >$	wCFds&wURF

3.4.4   Two-cell functional fault models

The found FPs for 2PFs in the previous section are compiled into a set of FFMs. The same notation is used as that for 2PF1s; e.g., wCFds&wCFds denotes a 2PF based on two weak CFds faults. Table 3.10 presents all introduced FFMs for 2PFs; see also column 'FFM' in Table 3.7 though Table 3.9. The FFMs are explained below. Remember that the 2PF2s are divided into three types: 2PF2$_a$, 2PF2$_v$, and 2PF2$_{av}$.

The 2PF2$_a$ FFMs

The 2PF2$_a$ has the property that the fault is sensitized in the v-cell by applying two simultaneous operations to the same a-cell; see Figure 3.3. Such simultaneous

Table 3.10. List of 2PF2 FFMs; $x \in \{0, 1\}$ and $op \in \{0r0, 1r1, 0w0, 0w1, 1w0, 1w1\}$

#	FFM	Fault primitives
1	wCFds&wCFds	$< 0wx : 0r0; 0/1/- >$, $< 0wx : 0r0; 1/0/- >$, $< 1wx : 1r1; 0/1/- >$, $< 1wx : 1r1; 1/0/- >$, $< rx : rx; 0/1/- >$, $< rx : rx; 1/0/- >$
2	wCFud&wCFud	$< 0wx : 0r0; 0/?/- >$, $< 0wx : 0r0; 1/?/- >$, $< 1wx : 1r1; 0/?/- >$, $< 1wx : 1r1; 1/?/- >$, $< rx : rx; 0/?/- >$, $< rx : rx; 1/?/- >$
3	wCFrd&wRDF	$< x; 0r0 : 0r0/1/1 >$, $< x; 1r1 : 1r1/0/0 >$
4	wCFdrd&wDRDF	$< x; 0r0 : 0r0/1/0 >$, $< x; 1r1 : 1r1/0/1 >$
5	wCFrrd&wRRDF	$< x; 0r0 : 0r0/1/? >$, $< x; 1r1 : 1r1/0/? >$
6	wCFir&wIRF	$< x; 0r0 : 0r0/0/1 >$, $< x; 1r1 : 1r1/1/0 >$
7	wCFrr&wRRF	$< x; 0r0 : 0r0/0/? >$, $< x; 1r1 : 1r1/1/? >$
8	wCFur&wURF	$< x; 0r0 : 0r0/?/0 >$, $< x; 0r0 : 0r0/?/1 >$, $< x; 0r0 : 0r0/?/? >$, $< x; 1r1 : 1r1/?/1 >$, $< x; 1r1 : 1r1/?/0 >$, $< x; 1r1 : 1r1/?/? >$
9	wCFrd&wWDF	$< x; 0r0 : 0w0/1/- >$, $< x; 1r1 : 1w1/0/- >$
10	wCFrd&wTF	$< x; 0r0 : 0w1/0/- >$, $< x; 1r1 : 1w0/1/- >$
11	wCFur&wUWF	$< x; 0r0 : 0w0/?/- >$, $< x; 1r1 : 1w1/?/- >$, $< x; 0r0 : 0w1/?/- >$, $< x; 1r1 : 1w0/?/- >$
12	wCFds&wTF	$< op : 0w1/0/- >$, $< op : 1w0/1/- >$
13	wCFds&wWDF	$< op : 0w0/1/- >$, $< op : 1w1/0/- >$
14	wCFds&wRDF	$< op : 0r0/1/1 >$, $< op : 1r1/0/0 >$
15	wCFds&wDRDF	$< op : 0r0/1/0 >$, $< op : 1r1/0/1 >$
16	wCFds&wRRDF	$< op : 0r0/1/? >$, $< op : 1r1/0/? >$
17	wCFds&wIRF	$< op : 0r0/0/1 >$, $< op : 1r1/1/0 >$
18	wCFds&wRRF	$< op : 0r0/0/? >$, $< op : 1r1/1/? >$
19	wCFds&wUWF	$< op : 0w0/?/- >$, $< op : 0w1/?/- >$, $< op : 1w1/?/- >$, $< op : 1w0/?/- >$
20	wCFds&wURF	$< op : 0r0/?/0 >$, $< op : 0r0/?/1 >$, $< op : 0r0/?/? >$, $< op : 1r1/?/0 >$, $< op : 1r1/?/1 >$, $< op : 1r1/?/? >$

operations are the allowed operations in the considered 2P memory; see Section 3.4.1. The 2PF2$_a$ consists of two FFMs; see Table 3.10:

1. wCFds&wCFds: Applying two allowed simultaneous operations to the a-cell will cause the v-cell to flip. This FFM consists of 12 FPs for the considered 2P memory.

2. wCFud&wCFud: Applying two allowed simultaneous operations to the a-cell forces the v-cell to an undefined state. This FFM consists also of 12 FPs for the considered 2P memory.

The 2PF2$_v$ FFMs

The 2PF2$_v$ has the property that the fault is sensitized in the v-cell by applying two simultaneous operations to the same v-cell, while the a-cell has to be in a certain state; see Figure 3.3. It consists of the following FFMs; see Table 3.10:

1. wCFrd&wRDF: Applying two simultaneous read operations to the v-cell will

cause the v-cell to flip, if the a-cell is in a certain state. The read operations then return *wrong* values. This FFM consists of four FPs.

2. wCFdrd&wDRDF: Applying two simultaneous read operations to the v-cell will cause the v-cell to flip, if the a-cell is in a certain state. The read operations return *correct* values. This FFM consists of 4 FPs.

3. wCFrrd&wRRDF: Applying two simultaneous read operations to the v-cell will cause the v-cell to flip, if the a-cell is in a certain state. The read operations then return *random* values. This FFM consists of 4 FPs.

4. wCFir&wIRF: Applying two simultaneous read operations to the v-cell return *wrong* values, if the a-cell is in a certain state. The v-cell maintains its correct stored value. This FFM consists of 4 FPs.

5. wCFrr&wRRF: Applying two simultaneous read operations to the v-cell return *random* values, if the a-cell is in a certain state. The v-cell maintains its correct stored value. This FFM consists of 4 FPs.

6. wCFur&wURF: Applying two simultaneous read operations to the v-cell forces the v-cell to an *undefined* state, if the a-cell is in a certain state. The returned read data values can be correct, wrong, or random. This FFM consists of 12 FPs.

7. wCFrd&wWDF: Applying simultaneously a read and a non-transition write to the v-cell causes the v-cell to flip, if the a-cell is in a certain state. This FFM consists of 4 FPs.

8. wCFrd&wTF: A transition write to the v-cell fails when a read operation is applied to the v-cell simultaneously, if the a-cell is in a certain state. This FFM consists of 4 FPs.

9. wCFur&wUWF: Applying simultaneously a read and a write to the v-cell brings the v-cell in an *undefined* state, if the a-cell is in a certain state. This FFM consists of 8 FPs.

The 2PF2$_{av}$ FFMs

The 2PF2$_{av}$ has the property that the fault is sensitized in the v-cell by applying two simultaneous operations: one to the a-cell and one to the v-cell; see Figure 3.3. It consists of the following FFMs; see Table 3.10.

1. wCFds&wTF: The v-cell fails to undergo a transition ($0 \rightarrow 1$ or $1 \rightarrow 0$) when it is written, *if* an operation is applied to the a-cell *simultaneously*. This FFM consists of 12 FPs.

2. wCFds&wWDF: A non-transition write operation ($0w0$ or $1w1$) performed to the v-cell causes a transition in the v-cell, *if* an operation is applied to the a-cell *simultaneously*. This FFM consists of 12 FPs.

3. wCFds&wRDF: A read operation performed on the v-cell changes the data in the v-cell, and returns an *incorrect* value on the output, *if* an operation is applied to the a-cell *simultaneously*. This FFM consists of 12 FPs.

4. wCFds&wDRDF: A read operation performed on the v-cell changes the data in the v-cell, and returns a *correct* value on the output, *if* an operation is applied to the a-cell *simultaneously*. This FFM consists of 12 FPs.

5. wCFds&wRRDF: A read operation performed on the v-cell changes the data in the v-cell, and returns a *random* value on the output, *if* an operation is applied to the a-cell *simultaneously*. This FFM consists of 12 FPs.

6. wCFds&wIRF: A read operation performed on the v-cell returns the incorrect logic value while keeping the correct stored value in the v-cell, *if* an operation is applied to the a-cell *simultaneously*. This FFM consists of 12 FPs.

7. wCFds&wRRF: A read operation performed on the v-cell changes the data in the v-cell, and returns a *random* value on the output, *if* an operation is applied to the a-cell *simultaneously*. This FFM consists of 12 FPs.

8. wCFds&wUWF: The v-cell is brought in an undefined state by a write operation, *if* an operation is applied to the a-cell *simultaneously*. This FFM consists of 24 FPs.

9. wCFds&wURF: The cell is brought in an undefined state by a read operation, *if* an operation is applied to the a-cell *simultaneously*. The returned data value by this read operation can be correct, wrong, or random. This FFM consists of 36 FPs.

3.4.5 Three-cell fault primitives

Two-port, three-cell fault primitives (2PF3s) involve three cells: two different a-cells and a v-cell; see Figure 3.3. The fault is sensitized when two simultaneous operations are applied to the two different a-cells. To denote the 2PF3s, the following FP notation will be used:

$< S_{a1} : S_{a2}; S_v/F/R >_{a1,a2,v}$. It denotes an FP whereby the sensitizing operations S_{a1} and S_{a2} are applied simultaneously to the a-cells c_{a1} and c_{a2}, respectively. S_v describes the state of the v-cell; i.e., $S_v \in \{0,1\}$. Note that R will be replaced with '$-$' since S_v can not be a read operation.

Table 3.11 gives all possible 2PF3 FPs; they consist in total of 144 FPs. Since S_{a1} and S_{a2} can only be sensitizing *operations* (i.e., one of the following operations: $\{0r0, 1r1, 0w0, 1w1, 0w1, 1w0\}$), there are $6\times6 = 36$ possible simultaneous operations '$S_{a1} : S_{a2}$'. In addition, $S_v \in \{0, 1\}$:

- If $S_v = 0$, then $F \in \{1, ?\}$ and $R = -$; the notation $< S_{a1} : S_{a2}; S_v/F/R >_{a1,a2,v}$ describes $36 \times 2 = 72$ FPs, since '$S_{a1} : S_{a2}$' can take on one of the 36 possible simultaneous operations while F can take on one of the two possible fault effects (FP1 through FP8 in Table 3.11).

- If $S_v = 1$, then $F \in \{0, ?\}$ and $R = -$; the notation $< S_{a1} : S_{a2}; S_v/F/R >_{a1,a2,v}$ describes $36 \times 2 = 72$ FPs (FP9 through FP16 in the Table 3.11).

Table 3.11. The complete set of 2PF3 FPs; $x, y, z, t \in \{0, 1\}$

#	$S_{a1} : S_{a2}$	S_v	F	R	$< S_{a1} : S_{a2}; S_v/F/R >$	FFM
1	$xrx : yry$	0	1	–	$< xrx : yry; 0/1/- >$	wCFds&wCFds
2	$xrx : yry$	0	?	–	$< xrx : yry; 0/?/- >$	wCFud&wCFud
3	$xrx : ywz$	0	1	–	$< xrx : ywz; 0/1/- >$	wCFds&wCFds
4	$xrx : ywz$	0	?	–	$< xrx : ywz; 0/?/- >$	wCFud&wCFud
5	$ywz : xrx$	0	1	–	$< ywz : xrx; 0/1/- >$	wCFds&wCFds
6	$ywz : xrx$	0	?	–	$< ywz : xrx; 0/?/- >$	wCFud&wCFud
7	$xwt : ywz$	0	1	–	$< xwt : ywz; 0/1/- >$	wCFds&wCFds
8	$xwt : ywz$	0	?	–	$< xwt : ywz; 0/?/- >$	wCFud&wCFud
9	$xrx : yry$	1	0	–	$< xrx : yry; 1/0/- >$	wCFds&wCFds
10	$xrx : yry$	1	?	–	$< xrx : yry; 1/?/- >$	wCFud&wCFud
11	$xrx : ywz$	1	0	–	$< xrx : ywz; 1/0/- >$	wCFds&wCFds
12	$xrx : ywz$	1	?	–	$< xrx : ywz; 1/?/- >$	wCFud&wCFud
13	$ywz : xrx$	1	0	–	$< ywz : xrx; 1/0/- >$	wCFds&wCFds
14	$ywz : xrx$	1	?	–	$< ywz : xrx; 1/?/- >$	wCFud&wCFud
15	$xwt : ywz$	1	0	–	$< xwt : ywz; 0/0/- >$	wCFds&wCFds
16	$xwt : ywz$	1	?	–	$< xwt : ywz; 0/?/- >$	wCFud&wCFud

3.4.6 Three-cell functional fault models

The total 144 two-port, three cell FPs (2PF3s) can be grouped into two FFMs.

1. wCFds&wCFds: Applying two simultaneous operations to two different a-cells will cause the v-cell to flip. This FFM consists of 74 FPs; see FPs $\#(2i - 1)$, whereby $1 \leq i \leq 8$, in Table 3.11.

2. wCFud&wCFud: Applying two simultaneous operations to two different a-cells forces the v-cell to an undefined state. This FFM consists also of 74 FPs; see FPs $\#2i$, whereby $1 \leq i \leq 8$, in Table 3.11.

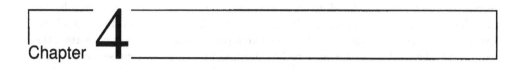

Chapter 4

Preparation for circuit simulation

4.1 Selected multi-port SRAM cell

4.2 Modeling of spot defects

4.3 Definition and location of opens

4.4 Definition and location of shorts

4.5 Definition and location of bridges

4.6 Simulation model

4.7 Simulation methodology

4.8 Simulation results for the fault free case

In the previous chapter, the complete space of memory faults has been introduced. These faults have a theoretical origin. In order to investigate their validity, an experimental and/or industrial analysis is required; e.g., defect injection in the memory cell and SPICE simulation.

This chapter discusses the preparation needed to begin with simulation in the next stages. The preparation includes selecting the circuit to be simulated, defining the complete set of defects that should be targeted, building an accurate simulation model, and an effective simulation methodology. Section 4.1 presents the structure of the multi-port memory cell considered for the analysis. Section 4.2 defines the spot defects, modeled as opens, shorts, and bridges. Section 4.3, Section 4.4, and Section 4.5 give all possible opens, shorts and bridges, respectively. Section 4.6 introduces the simulation model. Section 4.7 represents the simulation methodology; while Section 4.8 gives simulation results for the defect free case.

4.1   Selected multi-port SRAM cell

Figure 4.1 shows a selected multi-port (MP) SRAM cell; the *differential p-port (pP) memory cell*. It has the advantage of good performance, since differential sense amplifiers can be used. Therefore, it is most popular in the industry; especially for embedded caches in microprocessor products. The cell is designed with the usual two invertors and two pass transistors for each extra port. In order to save chip area, the pull-up transistors L_T and L_F have to be minimized in area. As a consequence, they lack a strong driving capability; so, the precharge of bit lines is required to speed up the read operations.

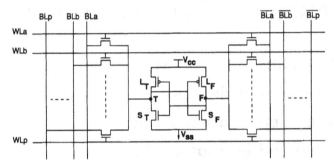

Figure 4.1. A differential p-port memory cell

4.2   Modeling of spot defects

Many faults in memory circuits are caused by undesired particles called *spot defects (SDs)*. SDs are geometrical features that emerge during the manufacturing process that were not originally defined by the integrated circuit (for instance, the memory) layout [22]. A SD is thus a randomly occurring region of extra or missing material in, or between, the layers used in the fabrication process. Since the electrical simulation will be considered, the physical SDs should be modeled electrically. The missing material will be modeled as disconnections, while the extra material will be modeled as undesired connections. These undesired disconnections and connections can be electrically divided into three groups:

- *An open*: which is an extra resistance within a connection. The resistor value called R_{op} is given by $0 < R_{op} \leq \infty$.

- *A short*: which is an undesired resistive path between a node and V_{cc} or V_{ss}. The resistor value called R_{sh} is given by $0 \leq R_{sh} < \infty$.

- *A bridge*: which is a parallel resistance between two connections, which are both different from V_{cc} and V_{ss}. The resistor value called R_{br} is given by $0 \leq R_{br} < \infty$.

It is important to note that the used model for SDs is simple; however it covers the majority of SDs. There are indeed defects that can not be modeled with a resistor; they require more complicated models.

From now on, the term SD will be used to mention an open, a short or a bridge.

SDs can occur in any sub-circuit of the memory circuit; i.e., the memory cell array, the address decoder, and the read/write logic. In this chapter, we will restrict ourself to SDs in the memory cell array. Figure 4.2 gives an overview of *Memory Cell Array Spot Defects (MCASDs)*. It should be noted that the discussion of SDs in this section will be done for a memory cell with any number of ports p.

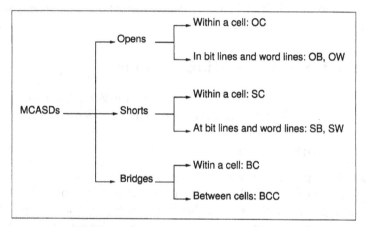

Figure 4.2. Classification of MCASDs

Many SDs can be identified in a pP memory. However, due to the symmetric structure of the cell, only a subset needs to be simulated. For identifying the not-to-be simulated SDs, the following terminology will be introduced [91]:

- *Similar behavior:* A spot defect SD1 (i.e., open, short or bridge) shows a similar behavior to SD2 if SD1 and SD2 present the same defect, but belong to different ports. E.g., a bridge between bit line BL_a and word line WL_b has a similar behavior to a bridge between bit line BL_b and word line WL_a; see Figure 4.1.

- *Complementary behavior:* SD1 shows a complementary behavior to SD2 if SD1 and SD2 present defects which locations in the memory cell are symmetrical with respect to each other; e.g., a bridge between the bit lines BL_a and BL_b at the true side has a complementary behavior to a bridge between $\overline{BL}_a$ and $\overline{BL}_b$ at the false side. In this case the functional fault behavior of SD1 is identical to that of SD2, with the only difference that all 1's are replaced with 0's and vice versa. E.g., if due to the presence of SD1 the operation read 0 (0r0) causes an up transition in the cell, then in the presence of SD2 the 1r1 operation causes a down-transition in the cell.

- *Interchanged behavior:* SD1 (involving two cells) shows an interchanged behavior to a SD2 (involving the same two cells) if the fault behavior of SD1 is identical to that that of SD2, with the only difference that the *aggressor cell* and the *victim cell* are interchanged; whereby the victim cell is the cell where the fault appears, while the aggressor cell is the cell to which the sensitizing operation (state) should be applied in order to sensitize a fault.

- *Interchanged complementary behavior:* SD1 shows an interchanged complementary behavior to SD2 if SD1 shows a complementary and interchanged behavior to SD2.

4.3   Definition and location of opens

Opens in the memory cell can be classified as *opens within a cell* (denoted as *OC*), and *opens at bit lines (OB)* and at *word lines (OW)*; see Figure 4.2.

4.3.1   Opens within a cell

In this case, the p-port memory cell will be considered without bit lines and word lines to which it is connected. In order to define all possible opens, the cell will be considered as a graph in which all branches can show such a defect. Figure 4.3 shows all possible locations of opens within a memory cell. Note that cells that belong to adjacent rows share the same V_{cc} or V_{ss} line, and that the opens at such lines are considered as opens within a cell. Opens at locations OCx and $OCxc$ will show *complementary* fault behaviors due to the symmetric structure of the memory cell, while opens at locations OCx and $OCxs$ will show *similar* fault behaviors due to the fact that the cell has p similar ports. For that reason, one can be limited to simulate the opens OCx only. From these, the behavior of the opens $OCxc$ and $OCxs$ can be derived. The first block of Table 4.1 shows the OCs. The first column lists the OCx opens, which are the minimal set that needs to be simulated, the third column gives the number of opens within one group. A *group* is a set of defects having a similar and/or a complementary fault behavior (e.g., opens at the source of the pull-up transistor at the true side consist of two opens OC1 and OC1c). Note that the total number of opens within a cell (including opens at V_{cc} and opens at V_{ss}) is $20 + 6p$, whereby p is the number of ports; note also that each port will add 6 possible opens to the list of opens (e.g., OC9, OC10, OC11, OC9c, OC10c, and OC11c); see Figure 4.3.

The fourth column in Table 4.1 classifies the opens into *Single-port Fault Defects (SFDs)* and *Multi-port Fault Defects (MFDs)*. The SFDs are spot defects that *only* can cause *single-port (SP)* faults; they can *not* cause faults typical for MP memories.

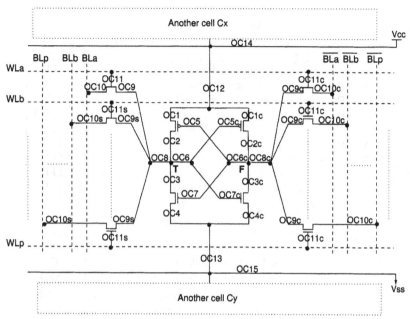

Figure 4.3. Opens within a cell

The SP faults are faults that can be sensitized using a single port. The MFDs are defects that can cause SP faults as well as *multi-port (MP)* faults; such faults require the use of multiple ports simultaneously in order to be sensitized. This classification is based on the simulation results done for a differential two-port (2P) memory [29, 31] (see Chapter 5): the SDs causing only SP faults in 2P memories are considered as SFDs, while SDs causing SP faults as well as *two-port* faults (i.e., faults requiring the use of two ports simultaneously in order to be sensitized) are considered as MFDs. Note that only OC3 and OC4 from Table 4.1 are MFDs. The fact that these two opens are MPDs can be explained as follows: When a single read operation is performed to the cell, in the presence of OC3 (or OC4), a resistor divider will be formed by the pass transistor and the pull down transistor together with OC3. If the resistance value of the defect is high enough, the voltage of the cell's node (during a read operation) will increase above the threshold voltage, and consequently the cell will flip. If two (or more) simultaneous read operations are performed, then two voltage dividers will be formed which have an additive effect on the cell's node. That means that the voltage on the node, in the presence of OC3, depends on the number of simultaneous read operations. Therefore OC3 is an MFD. Note that given the results for a 2P memory, then in order to examine a p-port memory, only the two MFDs (OC3 and OC4) need to be re-simulated. This eliminates a lot of simulation work.

Table 4.1. List of opens

Open	Description	#	Class
OC1	Source of pull-up at true side broken	2	SFD
OC2	Drain of pull-up at true side broken	2	SFD
OC3	Drain of pull-down at true side broken	2	**MFD**
OC4	Source of pull-down at true side broken	2	**MFD**
OC5	Gate of pull-up at true side broken	2	SFD
OC6	Cross coupling at true side broken	2	SFD
OC7	Gate of pull-down at true side broken	2	SFD
OC8	The connection of two pass transistors to T broken	2	SFD
OC9	Pass transistor connection to T broken	$2p$	SFD
OC10	Pass transistor connection to bit line broken	$2p$	SFD
OC11	Gate of pass transistor at true side broken	$2p$	SFD
OC12	V_{cc} path of the cell broken	1	SFD
OC13	V_{ss} path of the cell broken	1	SFD
OC14	V_{cc} path shared by adjacent cells broken	1	SFD
OC15	V_{ss} path shared by adjacent cells broken	1	SFD
OB_w	the bit line BL_i at the write side broken	$2p$	SFD
OB_r	the bit line BL_i at the read side broken	$2p$	SFD
OW	the word line WL_i broken	p	SFD

4.3.2   Opens at bit lines and word lines

Bit lines and word lines are connected to many cells. Therefore, an open at a bit line or a word line can influence the behavior of the operations applied to all these cells. In the following, first opens at bit lines will be discussed and thereafter at word lines.

Opens at bit lines: If we consider that the memory cell array is located between the read and the write circuit, then the opens at bit lines can occur in the following locations:

- An open between the cell and the write circuits (denoted as OB_w).
- An open between the cell and the read circuits (denoted as OB_r).

Since there are p pairs of bit lines connected to each cell, $4p$ opens at bit lines can exist; $2p$ opens at the side of the write circuits and $2p$ opens at the side of the read circuits. However, one only needs to simulate two opens (e.g., OB_w and OB_r at BL_i), because the behavior of the other opens (e.g, opens at $\overline{BL_i}$, BL_j, $\overline{BL_j}$, etc.) can be derived (i and j denote any two different ports). This is because opens at bit lines belonging to different ports and to the true side (e.g., BL_i and BL_j) have similar behaviors, while opens at the false side (e.g., $\overline{BL_i}$ and $\overline{BL_j}$) have complementary behaviors to opens at the true side.

Opens at word lines: The word lines are only driven by the row decoder. Since the opens at the pass transistor gates have already been defined as opens within a cell, the only remaining opens are those in the common word lines. The influence of such opens is the same for all cells along the word lines. We will define OW as an open at the word line WL_i; note that the total number of opens at word lines is p, and that they all have similar behaviors (e.g., an open at WL_j has a similar behavior to an open at WL_i).

The second block in Table 4.1 lists the OBs and the OWs; the minimal set of opens at bit lines and at word lines that has to be simulated consists of three opens while there are $5p$ possible opens at bit lines and word lines. Note that all these opens can cause only SP faults and no special faults for MP memories since the fault effects of such defects can only impact the operation applied via the port to which the SD belongs [29, 31]; e.g., an open at the word line of port i (P_i) can only impact the operations performed via P_i.

4.4   Definition and location of shorts

The shorts are classified as *shorts within a cell* (denoted as *SC*) and *shorts at bit lines (SB)* and at *word lines (SW)*; see Figure 4.2. Power shorts (i.e., shorts between V_{cc} and V_{ss}) are excluded, since they do not belong to the class of memory cell array faults; they impact the behavior of the whole circuit.

4.4.1   Shorts within a cell

To define shorts within a cell (SCs), a cell has to be considered as a graph in which all nodes can show a short. The cell is considered without bit lines and word lines. Each short is defined as a pair of nodes in which one node is V_{cc} or V_{ss}. The first block of Table 4.2 lists the possible SCs; shorts at F show complementary (Comp.) behaviors to shorts at T; see Figure 4.1. Note that the number of shorts within a cell is 4, irrespective of the number of ports p. In addition, and based on the simulation results of 2P memories [29, 31], SC1 can only cause SP faults, while SC2 can cause SP faults as well as special MP faults. Note that in the presence of SC2, a voltage divider will be formed during the read operations; the fault effect is then similar to that of OC3 and OC4.

4.4.2   Shorts at bit lines and at word lines

The cells belonging to the same column or the same row are connected to the same bit lines and word lines, respectively. Therefore, shorts at bit lines (SBs) and at word lines (SWs) can affect the behavior of all operations performed to these cells. Shorts at bit line BL_i and at word line WL_i have similar behaviors to shorts at BL_j and at WL_j, respectively; and shorts at $\overline{BL_i}$ have complementary behaviors to shorts at

Table 4.2. List of shorts

Shorts		Comp. behavior	#	Class
SC1	T-V_{cc}	F-V_{cc}	2	SFD
SC2	T-V_{ss}	F-V_{ss}	2	**MFD**
SB1	BL_i-V_{cc}	$\overline{BL_i}$-V_{cc}	$2p$	SFD
SB2	BL_i-V_{ss}	$\overline{BL_i}$-V_{ss}	$2p$	SFD
SW1	WL_i-V_{cc}		p	SFD
SW2	WL_i-V_{ss}		p	SFD

BL_i, whereby i and j can be any two different ports. The second and the third block of Table 4.2 list the possible SBs and SWs; shorts with complementary behavior are grouped together in the same row. The number of shorts within each group is also given in the table. The total number of SBs and SWs for a MP cell with p ports is $6p$; while one needs to simulate only four. Note that all SBs and SWs are SFDs, since the fault effects of such defects can only impact the operation applied via the port to which the SD belongs [29, 31].

4.5 Definition and location of bridges

A bridge in a p-port memory cell array can connect any arbitrary pair of nodes. However, the following assumptions are made:

1. The nodes have to be located close to each other; therefore, the bridge can occur only within a singe cell or between physically adjacent cells.

2. The defect can involve two nodes *at the most*.

These two assumptions are verified, based on the real data found using *Inductive Fault Analysis (IFA)*, which shows that the occurrence probability of defects involving more than two nodes is very small (3.4% on the average), since they require that a defect has to be very large [29, 31]. The bridges in the memory cell array can be divided into two groups:

- *Bridges within a cell (BCs):* All bridges connecting two nodes of the same cell, including the p pairs of bit lines and the p word lines to which it is connected. Bridges whereby each of the two connections are bit lines, word lines, or one bit line and one word line are also included.

- *Bridges between cells (BCCs):* All bridges connecting nodes of adjacent cells, including the bit lines and the word lines to which the cells are connected. Bridges whereby the two connections are both bit lines, word lines or one bit line and one word line are also included.

4.5.1 Bridges within a cell

To define all possible *bridges within a cell (BCs)*, the cell has to be considered as a graph in which each node can be connected to another by a bridge. Each p-port cell consists of $3p + 2$ nodes n_i ($n_i \in \{T, F, BL_i, \overline{BL}_i, WL_i\}$): a true and a false node (T,F), $2p$ bit lines ($BL_i, \overline{BL}_i$), and p word lines (WL_i); whereby i is one of the p ports ($i \in \{a, b, c, ..., p\}$). Therefore, there are $C_2^{3p+2} = \frac{(3p+2)!}{2!(3p)!} = \frac{9p^2+9p+2}{2}$ bridges. Table 4.3 shows all possible bridges within a cell. Note that bridges with similar or complementary behavior are grouped together in the same row, such that one can restrict the simulation to only one bridge of each row; e.g., only the first column of the table. The total number of bridges within one group is given in the third column of the table; the class of BCs is also given, based on the simulation results for 2P memories [29, 31]. Note that only BC6 and BC7 can cause special faults in MP memories since they involve bit lines belonging to different ports.

Table 4.3. List of bridges within a cell (BCs); $i, j \in \{a, b, ..., p\}$

Bridge		Comp. behavior	#	Class
BC1	T-F		1	SFD
BC2	T-BL_i	F-$\overline{BL}_i$	$2p$	SFD
BC3	T-$\overline{BL}_i$	F-BL_i	$2p$	SFD
BC4	T-WL_i	F-WL_i	$2p$	SFD
BC5	BL_i-$\overline{BL}_i$		p	SFD
BC6	BL_i-BL_j	$\overline{BL}_i$-$\overline{BL}_j$	$p(p-1)$	**MFD**
BC7	BL_i-$\overline{BL}_j$		$p(p-1)$	**MFD**
BC8	BL_i-WL_i	$\overline{BL}_i$-WL_i	$2p$	SFD
BC9	BL_i-WL_j	$\overline{BL}_i$-WL_j	$2p(p-1)$	SFD
BC10	WL_i-WL_j		$\frac{p(p-1)}{2}$	SFD

4.5.2 Bridges between cells

Bridges *between cells (BCCs)* consist of BCCs in the *same row (rBCCs)*, BCCs in the *same column (cBCCs)*, and BCCs on the *same diagonal (dBCCs)*. To establish all possible BCCs, the configuration shown in Figure 4.4 will be considered. It consists of four cells; namely C_1, C_2, C_3 and C_4. Note that the adjacent cells can belong to the same column, the same row, or to the same diagonal. The cells C_1 and C_3, as well as the cells C_2 and C_4, are adjacent in the same row and therefore have common word lines; while the cells C_1 and C_2 (as well as the cells C_3 and C_4) are adjacent in the same column and therefore have common bit lines.

1. Bridges between cells in same row

In order to find all possible bridges between adjacent cells in the same row (rBCCs), only C_1 and C_3 have to be considered; see Figure 4.4. Both C_1 and C_3 consist of

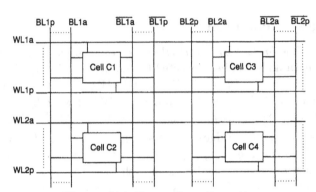

Figure 4.4. Four cell configuration

$2 + 3p$ nodes: C_1 consists of T1, F1, $BL1_i$, $\overline{BL1}_i$, and $WL1_i$, while C_3 consists of T3, F3, $BL2_i$, $\overline{BL2}_i$, and $WL1_i$; $i \in \{a, b, ..., p\}$. Since both cells have a common word line, only the true or false node (T1, F1) of C_1 and its p pairs of bit lines can form a bridge with the true/false node or with the p pairs of bit lines of C_3. Therefore, there are $(2 + 2p) * (2 + 2p) = 4p^2 + 8p + 4$ possible bridges, $n_1 - n_2$, between the two cells; whereby n_1 is a node of C_1 ($n_1 \in \{T1, F1, BL1_i, \overline{BL1}_i, \}$) and n_2 is a node of C_3 ($n_2 \in \{T3, F3, BL2_j, \overline{BL2}_j\}$). The first block of Table 4.4 shows all possible rBCCs; port i and j indicate any two *different* ports. Bridges with a complementary, an interchanged (Inter.), or an interchanged complementary (I.C.) behavior are grouped together in the same row. The total number of bridges within one group is given in the fifth column of the table. Note that the $4p^2 + 8p + 4$ possible bridges are grouped in only eight groups. The class of rBCCs is also given based on the simulation results found for 2P memories [29, 31]. Note that all rBCCs can cause special faults in MP memories.

2. Bridges between cells in same column

In order to find all possible bridges between adjacent cells in the same column (cBCCs), only C_1 and C_2 have to be considered; see Figure 4.4. Both C_1 and C_2 consist of $2 + 3p$ nodes: C_1 consists of T1, F1, $BL1_i$, $\overline{BL1}_i$, and $WL1_i$, while C_2 consists of T2, F2, $BL1_i$, $\overline{BL1}_i$, and $WL2_i$; $i \in \{a, b, ..., p\}$. Note that the two cells share the same bit lines. Therefore, there are $(2 + p) * (2 + p) = p^2 + 4p + 4$ possible bridges, $n_1 - n_2$, between C_1 and C_2; whereby $n_1 \in \{T1, F1, WL1_i\}$ and $n_2 \in \{T2, F2, WL2_i\}$. A bridge between the bit lines and the nodes T2 or F2 is excluded since it belongs to bridges within a cell, which are already considered in Section 4.5.1. The second block of Table 4.4 lists the $p^2 + 4p + 4$ possible cBCCs; they are grouped into five groups. Note that only three cBCC groups can cause special faults in MP memories.

Table 4.4. Bridges between adjacent cells (BCCs)

Bridge BCC		Comp. behavior	Inter. behavior	I.C. behavior	# of brid.	Class
rBCC1	T1-T3	F1-F3			2	MFD
rBCC2	T1-F3	F1-F3			2	MFD
rBCC3	T1-$BL2_i$	F1-$BL2_i$	$\overline{BL1}_i$-T3	$\overline{BL1}_i$-F3	4p	MFD
rBCC4	T1-$\overline{BL2}_i$	F1-$\overline{BL2}_i$	$\overline{BL1}_i$-T3	$BL1_i$-F3	4p	MFD
rBCC5	$BL1_i - BL2_i$	$BL1_i - BL2_i$			2p	MFD
rBCC6	$BL1_i - BL2_j$	$BL1_i - BL2_j$			2p(p-1)	MFD
rBCC7	$\overline{BL1}_i - \overline{BL2}_i$		$\overline{BL1}_i - BL2_i$		2p	MFD
rBCC8	$BL1_i - \overline{BL2}_j$		$\overline{BL1}_i - BL2_j$		2p(p-1)	MFD
cBCC1	T1-T2	F1-F2			2	MFD
cBCC2	T1-F2	F1-T2			2	MFD
cBCC3	T1-$WL2_i$	F1-$WL2_i$	T2-$WL1_i$	F2-$WL1_i$	4p	MFD
cBCC4	$WL1_i$-$WL2_i$				p	SFD
cBCC5	$WL1_i$-$WL2_j$				p(p-1)	SFD
dBCC1	T1-T4	F1-F4			2	MFD
dBCC2	T1-F4	F1-T4			2	MFD

3. Bridges between diagonal cells

The possible bridges between cells belonging to the same diagonal, dBCCs, (i.e., C_1 and C_4 of Figure 4.4) consist only of four bridges; see the third block of Table 4.4. All other bridges between the nodes of C_1 and the nodes of C_4 are already considered in rBCCs and cBCCs; this is because C_4 has the same word lines as C_2 and the same bit lines as C_3.

4.6   Simulation model

The SPICE based *Circuit Simulation Environment* [1] *(CSE)* has been used for the simulation. Since CSE requires too much simulation time for a complete memory, an appropriate simulation model has to be built, which will both accurately describe the behavior of the memory while only requiring a reasonable simulation time. The accuracy of the simulation model determines the accuracy of the results, which implies that the model has to approximate the actual memory structure as close as possible.

In the next stage, the simulation will be done for a two-port (2P) as well as for a three-port (3P) memory. Therefore, an appropriate simulation model has to be built for both cases. In this section, the model will be discussed only for a 2P memory; the model for 3P is built in a similar way.

The 2P memory simulation model is based on an Intel real cache design. The cache has a size of 16KBytes, and is divided into four sub-arrays; each contains

---

[1]The Intel internal electrical circuit simulator provides an unified graphics-based environment for the circuit design process, and it supports the complete simulation cycle from setup and execution to processing and viewing the results

4KBytes, arranged as 128 rows and 256 columns plus 32 columns for parity bits. Each sub-array can be considered as a stand alone unit; therefore, the simulation model can be build only for one sub-array.

In order to examine the fault effects between adjacent memory cells, a 2×2 memory cell array is required. However, a 3×3 memory cell array is used since it will be the subject of other simulations. Each cell can be accessed separately and independent of each other (i.e., the 3×3 memory cell array is *bit oriented*). The cells have to be chosen in this way, because the considered 2P memory has an *interleaved organization*; this means that adjacent cells contain data belonging to different data-words. Since each sub-array is arranged as 128 rows by 256 columns, plus 32 columns for parity bits, it is important to take the loading of cells that share the same word lines or bit lines as the 3×3 model into account. To do that, cells sharing the same word lines (i.e., 3×285), and cells sharing the same bit lines (i.e., 125×3) as the 3×3 memory cell model are added to the simulation model, as well as the dummy cells; see Figure 4.5. Note that the 3×3 model is inserted at the end of word lines, and in the middle of the bit lines. It should be noted that in order to not drive the word line with an ideal voltage source, word line drivers are added; they consist of two invertors.

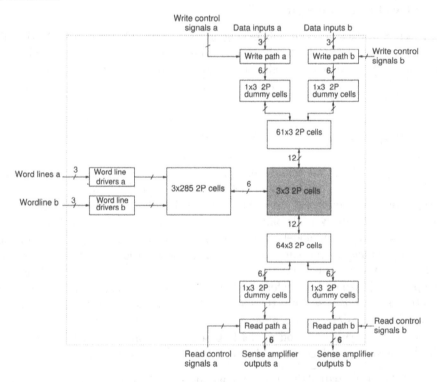

Figure 4.5. Simulation model for 2P memory

The cells sharing the same word lines as the 3×3 model (i.e., 3×285 blocks; see Figure 4.5) are further divided into five sub-blocks of 3×48, and one sub-block of 3×45. On the other hand, the cells sharing the same bit lines as the 3×3 model (i.e., the 61×3 and 64×3 blocks in Figure 4.5) are divided into 8 sub-blocks. The 61×3 block is divided into three 16×3 sub-blocks, and one 13×3 sub-block; while the 64×3 block is divided into four 16×3 sub-blocks. This is done in order to distribute the loading on the word line and on the bit lines. Moreover, the word lines and the bit lines interconnections are modeled using a *File Tracking Resistance Capacitor*[2] (FTRC); it is a pi-model formed by two capacitors and a resistor. This model includes the resistance of the interconnection as well as the coupling between the adjacent interconnections, and is inserted between each two sub-blocks of the loading cells. The three elements of the FTRC model (i.e., two capacitors and a resistance) are calculated based on the layout parameters of the design (i.e., width and the length of the bit lines and the word lines, the metal spacing, etc.)

The simulation model consists also of two duplicated read path circuits as well as two duplicated write path circuits. Such circuits allow the performance of the read and the write operations to the 3x3 memory cell array. The read path circuit, which is the route of the read data from the memory cell array to the data output lines, consists of bit line keepers, bit line precharge circuits, sense amplifier, multiplexers, and sense amplifiers. The write circuit, which is the route of the write data from data input to the memory cell array, consists of the bit line keepers, the bit line precharge circuits, the write multiplexer, and the write drivers. The timing of the input control signals for read operations and write operations specified by the designer is used for the simulation.

4.7 Simulation methodology

As mentioned in the previous section, the simulation will be done for 2P as well as for 3P memories. This section discusses the method to be used to run the simulation in the presence of a certain spot defect in the circuit. The methodology will be presented only for a 2P memory; the methodology for a 3P memory can be derived in a similar way.

The spot defects (SDs) discussed in Section 4.3 through Section 4.5 can be classified into *single-cell defects*, and *multi-cell defects*.

1. Single-cell defects: these are defects that only impact the behavior of a single cell; e.g., an open at the pull down transistor at the true side of the cell can impact only the behavior of this cell. These defects consist of opens within a cell, shorts within a cell, and bridges within a cell whereby no bit line and no word line is involved.

---

[2]Model used at Intel

2. Multi-cell defects: these consist of defects whereby the involved nodes belong to different cells, and defects at connections shared by more than one cell. Note that such defects can impact more than one cell; e.g., a bridge between two different cells can impact the behavior of both cells. These defects consist of opens at bit lines and word lines, shorts at bit lines and word lines, bridges within a cell whereby the bit lines or the word line are involved, and bridges between cells.

In the remainder of this section the simulation methodology for each of the two defect classes, single-cell and multi-cell defects, for a 2P memory will be discussed.

4.7.1 Single-cell defects

The simulation methodology for 2P memory has to examine all possible operations involving a single cell, applied to the defective *single* cell: single-port operations involving a single cell (denoted as 1P-Ops) as well as two-port operations involving a single cell (denoted as 2P-Op1s). 1P-Ops use only one port at a time (e.g., the 1P-Op: '0r0' performs a single read operation), while 2P-Op1s use the two ports simultaneously (e.g., the 2P-Op1: '0r0 : 0r0'performs two simultaneous read operations). Table 4.5 and Table 4.6 show all allowed 1P-Ops and 2P-Op1s in the to be analyzed 2P memory, respectively. Note that, depending on the state of the cell to which the operation is applied, two types of write operations can be distinguished: transition (e.g., '0w1') and non-transition (e.g., '0w0') write operations. Note also that two simultaneous write operations to the same cell are not allowed, while simultaneous reading and writing of the same location (e.g., '0r0 : 0w1') is allowed; however, the read data in this case will be discarded.

Table 4.5. 1P-Ops   **Table 4.6.** The 2P-Op1s

Oper.	State
0w0	0
0w1	0
1w0	1
1w1	1
1r1	1
0r0	0

Oper.	State
0r0 : 0r0	0
1r1 : 1r1	1
0r0 : 0w0	0
0r0 : 0w1	0
1r1 : 1w1	1
1r1 : 1w0	1

In order to reduce the simulation time, different *operation sequences* can be formed based on the 1P-Ops and 2P-Op1s of Table 4.5 and Table 4.6. These sequences can be used during the simulation, instead of running each operation separately. For instance the singe-port operation sequence '0w1, 1r1, 1w0, 0r0' describes a sequence of four operations applied in the given order. Running the sequence will require only to write one command file, rather than four if each operation of the sequence is verified separately.

Furthermore, it is important to examine the behavior of the cell for a long time in the presence of a certain time-dependent defect. Therefore, an additional simulation, the *data-retention simulation*, is needed. In this case, the cell will be initialized to a given state (0 or 1), and a wait time will be added to inspect if the cell is able to maintain its state.

The simulation methodology that will be used in order to simulate single-cell defects in a 2P memory cell is given below. Note that a strong fault is denoted using the fault primitive notation (see Chapter 3); while a weak[3] fault is denoted as 'wF'.

A. *For each single-cell defect, examine the resistance range from 0 to ∞ and do:*

    A.1. *Verify the correctness of 1P-Ops by simulating the proper sequence(s), and by inspecting the cell under simulation:*

- *If a strong fault occurs, then translate the behavior of the fault into a fault primitive, and go to Step A.3.*
- *If a weak fault occurs, then go to Step A.2.*
- *If no fault occurs, then go to Step A.2.*

    A.2. *Verify the correctness of 2P-Op1s by simulating the proper sequence(s) and by inspecting the cell under simulation:*

- *If a strong fault occurs, then translate the behavior of the fault into a fault primitive, and go to Step A.3.*
- *If a weak fault occurs, then denote the fault as wF and go to Step A.3.*
- *If no fault occurs, then go to Step A.3.*

    A.3. *END*

B. *For each single-cell defect, perform the data-retention simulation.*

It is of interest to know how the examination of the resistance range from 0 to ∞ has been done; this applies to single-cell as well as to multi-cell defects.

For the opens, the first simulation has been done for the case that $R_{op} = \infty$; while for shorts (bridges), the first simulation has been done for the case that $R_{sh} = 0$ ($R_{br} = 0$). If no fault occurs in these cases, then it does not make sense to simulate other cases for smaller values of R_{op} or larger values of R_{sh} (R_{br}). If a fault occurs, then the simulation has been repeated for smaller values of R_{op}, and bigger values of R_{sh} (R_{br}). It should be noted that in case the simulated open causes the gate of a certain transistor to be floating (for $R_{op} = \infty$), the simulation has been repeated

---

[3]the definitions of strong and weak faults are given in Section 3.4.2

for different initial gate voltages (between 0 and V_{cc}) in order to get all possible fault effects. For example the open OC7 causes the gate of the pull down transistor at the true side to be floating for $R_{op} = \infty$ (see Figure 4.3); therefore during the simulation, the initial voltage of the floating gate (V_f) is examined in the range 0-V_{cc}. The boundary between the different behaviors of the cell (i.e., proper operation, strong fault, weak fault) is searched by stepping through the resistor value range. This stepping starts for opens with $R_{op} = 10^k$ in which k is decreased by 1 each time (starting with $k = 15$); while for shorts (and bridges) the stepping starts with $R_{sh}(R_{br}) = 10^k$ in which k is increased by 1 each time (starting with $k = 0$). If a transition between two different behaviors (e.g., proper behavior and strong fault), then the region between 10^k and 10^{k-1} for opens is examined with a step size of 10^{k-1}; while the region between between 10^k and 10^{k+1} for shorts and bridges is examined with a step size of 10^k. These actions are repeated until the boundary has been found with the desired accuracy.

4.7.2 Multi-cell defects

These defects consist of opens at bit lines and word lines, shorts at bit lines and word lines, bridges within a cell whereby the bit line(s) or the word line is involved, and bridges between cells. Note that such defects can impact more than one cell; e.g., a bridge between two different cells can impact the behavior of both cells.

Table 4.7. The 2P-Op2s; $x, y \in \{0, 1\}$

Operation	State		Operation	state	
	c_1	c_2		c_1	c_2
$xw0_{c_1} : yw0_{c_2}$	x	y	$0r0_{c_1} : 0r0_{c_2}$	0	0
$xw0_{c_1} : yw1_{c_2}$	x	y	$0r0_{c_1} : 1r1_{c_2}$	0	1
$xw1_{c_1} : yw0_{c_2}$	x	y	$1r1_{c_1} : 0r0_{c_2}$	1	0
$xw1_{c_1} : yw1_{c_2}$	x	y	$1r1_{c_1} : 1r1_{c_2}$	1	1
$xw0_{c_1} : 0r0_{c_2}$	x	0	$0r0_{c_1} : yw0_{c_2}$	0	y
$xw0_{c_1} : 1r1_{c_2}$	x	1	$1r1_{c_1} : yw0_{c_2}$	1	y
$xw1_{c_1} : 0r0_{c_2}$	x	0	$0r0_{c_1} : yw1_{c_2}$	0	y
$xw1_{c_1} : 1r1_{c_2}$	x	1	$1r1_{c_1} : w1_{c_2}$	1	x

For the simulation of multi-cells defects, all allowed operations in the considered 2P memory have to be verified. These consist of operations involving a single cell (i.e, 1P-Ops and 2P-Op1s) and *two-port operations involving two cells* (denoted as *2P-Op2s*). The 1P-Ops and 2P-Op1s are shown in Table 4.5 and Table 4.6, respectively; while the 2P-Op2s are shown in Table 4.7. Note that the considered 2P memory allows the following 2P operations to be performed to different locations: simultaneous write operations (e.g., '$0w0_{c1} : 0w0_{c2}$' denotes two simultaneous non transition $w0$

operations to cells c_1 and c_2), simultaneous read operations, and simultaneous write and read operations. If we assume that the cells c_1 and c_2 are the cells involved in the defect, then the operation sequences involving a single-cell have to be performed twice: c_1 as accessed cell and c_2 as accessed cell. In addition, the different initial states of the cells c_1 and c_2 have to be taken into account.

The simulation methodology used to simulate multi-cell defects in the 2P memory cell is the following:

For each multi-cell defect, examine the resistance range from 0 to ∞ and do:

A. Verify the correctness of operations involving a single cell for the (two) cells involved in the defect:

    A.1. Verify the correctness of 1P-Ops by simulating the proper sequence(s), and by inspecting the cell under simulation and the neighbor cells:

        – *If a strong fault occurs, then translate the behavior of the fault into a fault primitive, and go to Step C.*

        – *If a weak fault occurs, then go to Step A.2.*

        – *If no fault occurs, then go to Step A.2.*

    A.2. Verify the correctness of 2P-Op1s by simulating the proper sequence(s) and by inspecting the cell under simulation and the neighbor cells:

        – *If a strong fault occurs, then translate the behavior of the fault into a fault primitive, and go to Step C.*

        – *If a weak fault occurs, then go to Step B.*

        – *If no fault occurs, then go to Step B.*

B. Verify the correctness of operations involving two cells:

    B.1. Verify the correctness of 2P-Op2s by simulating the proper sequence(s) and by inspecting the two cells under simulation (i.e., accessed cells) and the neighbor cells:

        – *If a strong fault occurs, then translate the behavior of the fault into a fault primitive, and go to Step C.*

        – *If a weak fault occurs, then denote the fault as wF and go to Step C.*

        – *If no fault occurs, then go to Step C.*

C. END

4.8 Simulation results for the fault free case

This section gives simulation results for a defect free 2P memory. After the simulation model has been built, and the command files have been written, the simulation has been run. Depending on the used parameters in the model, two types of simulations have been performed:

1. Simulation based on the *extracted* parameters from the layout. In this case loading cells in the model of Figure 4.5 are replaced during the simulation with data files extracted from the layout; as well as the FTRC (model used for bit lines and word lines connections).

2. Simulation based on the *estimated* parameters. In this case no extracted data has been used. The used parameters (node capacitance, resistance, etc.) are estimated based on the Intel schematic estimator (PAREST).

The simulation has been done for both cases. The results are visualized graphically. Figure 4.6 shows the simulation results using extracted parameters. The results obtained by simulating the single-port operation sequence '$0w1, 1r1, 1w0, 0r0$', applied via port P_a. The figure consists of three individual graphs; each with its own vertical and horizontal axises. At the right side of the graphs, for each wave a label is shown which defines the connection or the node at which the voltage is measured. In addition, a number 0 or 1 is attached to the label; this number is used in the wave to make a clear distinction between waves of different labels. For example in the top graph, node T (i.e., the true node of the accessed cell) has a number 0, while node F (i.e., the false node of the accessed cell) has a number 1.

The first graph gives the voltage of the two nodes of the cell (T and F). The second graph shows the voltage of the two bit lines BLa and $\overline{BLa}$ (denoted as $BLa\#$). The third graph presents the outputs of the sense amplifier which are denoted as $SAout$ and $SAout\#$.

The range of the time axis is $4T$, because four operations of duration T are performed. A first look at the waves of the figure shows that the waves before $t = 2T$ are similar to those after $t = 2T$. This is due to the symmetrical structure of the memory cell. The data in the cell is initialized to 0. At $t = 0$, the precharge phase starts by precharging the bit lines to V_{cc}. At $t = 0.40T$, the write multiplexer is activated and the $BLa\#$ is pulled low. Since the word line has been selected, the '1' is written in the cell, and consequently the state of the cell changes. At $t = 0.71T$, the write multiplexer is unactived and the cell maintains its state. At $t = 0.76T$, a new precharge phase starts. After this precharge phase, the word line is set high at $t = 1.29T$, and thereafter the read multiplexer is activated at $t = 1.67T$. Since the content of the cell is 1, the $BLa\#$ will be discharged until the voltage difference between BLa and $BLa\#$ becomes equal to a certain voltage ΔV required for the sense amplifier, which will then sense this difference and produce 1 as output. Note that due to the leakage/ capacitive coupling, the BLa is also discharged a little. This has no consequences, because the discharge of $BLa\#$ is very fast compared with that of BLa. The operations '$1w0, 0r0$' show the same waveforms; the only difference is that they show a complementary behavior of '$0w1, 1r1$'. It should be noted that the

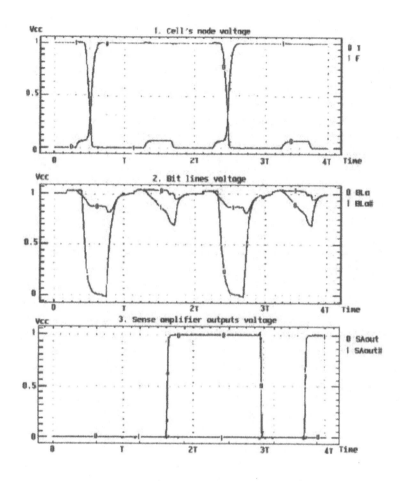

Figure 4.6. Simulation results of '$0w1, 1r1, 1w0, 0r0$' using extracted parameters

run time for the simulated operation sequence, using extracted parameters from the layout, is about 826.78s.

Figure 4.7 shows the simulation results for the case the estimated parameters were used. The results were obtained by simulating the same sequence as the above case (i.e., '$0w1, 1r1, 1w0, 0r0$'). The run time required in this case is 54.41s; which is more than 15 times less than the first case. Note that only a very tiny difference can be seen between the results of the two cases. Based on these two facts (i.e., run time and accuracy of simulation results), and due to the number of simulations that has to be performed, the simulation model based on the estimated parameters will be used from now on.

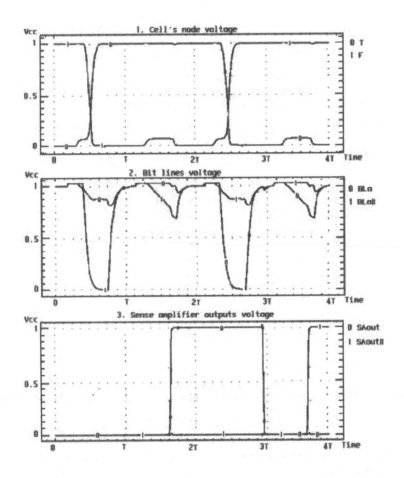

Figure 4.7. Simulation results of '$0w1, 1r1, 1w0, 0r0$' using estimated parameters

Part II

Testing single-port and two-port SRAMs

This part concerns with developing realistic fault models, tests, and a test strategies for two-port SRAMs. It is divided into three chapters. Chapter 5 introduces realistic fault models, together with their probabilities based on defect injection, SPICE simulation and Inductive Fault Analysis. Chapter 6 establishes tests for the introduced faults. It also gives an industrial evaluation of such tests. Chapter 7 covers the impact of port restrictions (e.g., read-only port, write-only port) on the faults models as well as on the tests.

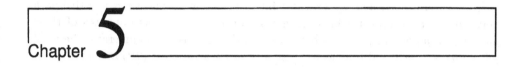

Chapter **5**

Experimental analysis of two-port SRAMs

5.1 The to-be simulated spot defects

5.2 Simulation results

5.3 Realistic fault models

5.4 Fault probability analysis

In Chapter 4, the preparation needed to begin with the multi-port SRAM simulation has been presented; this included a complete analysis of spot defects in a differential multi-port memory cell array, the simulation model, as well as the simulation methodology.

In this chapter, the simulation will be done for a dual-port memory. Section 5.1 derives the to be simulated modeled spot defects based on the analysis done in Chapter 4 for a p-port memory. Section 5.2 presents the electrical simulation results using the fault primitive notation. Section 5.3 transforms the electrical faults into realistic functional fault models. Section 5.4 gives the probability of occurrence of such faults based on two approaches; first by assuming that all modeled spot defects are equal likely to occur and thereafter by performing inductive fault analysis to two different layouts representing the same electrical circuit.

5.1   The to-be simulated spot defects

In Chapter 4, spot defects (SDs) are modeled as opens, shorts and bridges; whereby an open is an extra resistance within a connection, a short is an undesired resistive path between a node and V_{cc} or V_{ss}, while a bridge is an undesired resistive path between two connections, which are not V_{cc} or V_{ss}. A complete analysis of these models in a p-port memory has been performed. This section enumerates the total set of opens, shorts and bridges to be simulated for a two-port (2P) memory.

Figure 5.1 shows a differential 2P memory cell, with two *read-write* ports, that will be the subject of this chapter. The following operations are allowed to be performed:

- Single read and write operations through each port.
- Two simultaneous read operations to the same location, as well as to different locations.
- Two simultaneous write operations to different locations.
- Simultaneous read and write to different locations.
- Simultaneous read and write to the same location. However, in that case the read data will be discarded; i.e., the write operation has a higher priority.

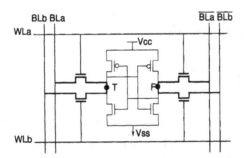

Figure 5.1. Differential two-port memory cell

In Chapter 4, all opens, shorts and bridges have been defined and located for a differential p-port memory; $p > 1$. For $p = 2$, it can be easily verified that there are 142 possible SDs: 42 opens, 16 shorts, and 84 bridges. For the bridges, the assumption is made that the nodes have to be located close to each other, such that bridges can occur only within a singe cell or between adjacent cells. The SPICE simulation of all opens, shorts and bridges will require a significantly long time due the total number of SDs that needs to be simulated. However, the 142 possible SDs in the 2P memory cell can be placed into 49 groups, whereby only one SD from each group needs to be simulated; the behavior of the other SDs within a group can be derived based on the simulated one. The grouping is based on the fact that the memory cell has a symmetrical structure and two similar ports; see Section 4.2.

Table 5.1. List of opens to be simulated for a 2P memory

Opens	Description
OC1	Source of pull-up at true side broken
OC2	Drain of pull-up at true side broken
OC3	Drain of pull-down at true side broken
OC4	Source of pull-down at true side broken
OC5	Gate of pull-up at true side broken
OC6	Cross coupling at true side broken
OC7	Gate of pull-down at true side broken
OC8	The connection of two pass transistors to T broken
OC9	Pass transistor connection to T broken
OC10	Pass transistor connection to bit line broken
OC11	Gate of pass transistor at true side broken
OC12	V_{cc} path of the cell broken
OC13	V_{ss} path of the cell broken
OC14	V_{cc} path of shared by adjacent cells broken
OC15	V_{ss} path of shared by adjacent cells broken
OB_w	The bit line BL_a at the write side broken
OB_r	The bit line BL_a at the read side broken
OW	The word line WL_a broken

Table 5.1 shows the minimal set of opens that needs to be simulated (see also Figure 5.1). They are divided into opens within a cell (OC), opens at bit lines (OB), and opens at word lines (OW).

The shorts are divided into shorts within a cell (SC), shorts at bit lines (SB) and at word lines (SW). The minimal set of shorts required for the simulation is shown in the first column of Table 5.2. Each short is defined as a pair of nodes in which one node is V_{cc} or V_{ss}.

On the other hand, the bridges have been divided into two groups:

• Bridges within a cell (BC): All bridges connecting two nodes of the same cell, including the two pairs of bit lines and the two word lines to which it is connected. The second column of Table 5.2 gives the minimal set of BCs required for the simulation; see also Figure 5.1.

• Bridges between cells (BCC): All bridges connecting nodes of adjacent cells, including the bit lines and the word lines to which the cells are connected. For establishing all possible BCCs, the configuration shown in Figure 5.2 has to be considered. It consists of four cells; namely, C_1, C_2, C_3 and C_4. Note that the adjacent cells can belong to the same column, the same row, or to the same diagonal. Therefore, the BCCs are further divided into BCCs between cells in the same *row* (rBCCs), BCCs between cells in the *column* (cBCCs), and BCCs between cells in the same *diagonal* (dBCCs). The third and the fourth column of Table 5.2 show the minimal set of BCCs that needs to be simulated. In the table, e.g., T1 (F1) denotes the true (false) node of cell C_1, see also Figure 5.2.

Figure 5.2. Four cell configuration for 2P memory

Table 5.2. Minimal set of shorts and bridges to be simulated for 2P memory

Shorts		Bridges	BCs	Bridges	rBCC	Bridges	c(d)BCCs
SC1	T-V_{cc}	BC1	T-F	rBCC1	T1-T3	cBCC1	T1-T2
SC2	T-V_{ss}	BC2	T-BL_a	rBCC2	T1-F3	cBCC2	T1-F2
SB1	BL_a-V_{cc}	BC3	T-$\overline{BL}_a$	rBCC3	T1-$BL2_a$	cBCC3	T1-$WL2_a$
SB2	BL_a-V_{ss}	BC4	T-WL_a	rBCC4	T1-$\overline{BL2}_a$	cBCC4	$WL1_a$-$WL2_a$
SW1	WL_a-V_{cc}	BC5	BL_a-$\overline{BL}_a$	rBCC5	$BL1_a - BL2_a$	cBCC5	$WL1_a$-$WL2_b$
SW2	WL_a-V_{ss}	BC6	BL_a-BL_b	rBCC6	$BL1_a - BL2_b$	dBCC1	T1-T4
		BC7	BL_a-$\overline{BL}_b$	rBCC7	$BL1_a - \overline{BL2}_a$	dBCC2	T1-F4
		BC8	BL_a-WL_a	rBCC8	$BL1_a - \overline{BL2}_b$		
		BC9	BL_a-WL_b				
		BC10	WL_a-WL_b				

5.2 Simulation results

The simulation has been done for all 49 SDs (see Table 5.1 and Table 5.2), by using the simulation methodology discussed in Chapter 4. Each electrical faulty behavior is reported in terms of a fault primitive (FP); if a strong fault is sensitized, then the FP notations introduced Chapter 3 is used to describe it. If a fault is only partially sensitized (e.g., a weak fault) then the fault is denoted as wF.

In order to save space, only simulation results for some SDs will be presented here; the results of all SDs are given in Appendix A. Table 5.3 lists the simulation results for two opens, four bridges within a cell, and three bridges between cells. The first column in the table gives the name of the simulated SD; see also Table 5.1 and Table 5.2. The second column gives the resistance regions[1] ordered in an increasing values,

---

[1] The exact resistance values for each region are process specific and Intel proprietary

the third and the fourth columns list the FP sensitized by the simulated SD and the derived complementary FP (if applicable), respectively; a FP is applicable if there is another modeled SD which has a fault behavior complementary to that of the SD simulated (see Section 4.3). The fifth column gives the class of the sensitized fault; i.e., single-port faults involving a single cell (1PF1s), single-port faults involving two cells (1PF2s), two-port faults involving a single cell (2PF1s), and two-port faults involving two cells (2PF2s). The table shows clearly that the sensitized fault is *strongly dependent* on the resistance value of the SD, and that some SDs cause only single-port faults, while other cause single-port faults as well as two-port faults; e.g., the OC3 can sensitize six different FPs:

- Region VI and V: In this case the cell flips during a $0r0$ operation, and the sense amplifier outputs are set to *incorrect* data (i.e., $< 0r0/1/1 >_v$) . That means that the fault is directly visible at the memory output. The fact that the cell flips during the $0r0$ operation can be explained as follows. Before a cell is read, the bit lines are precharged to V_{cc}; see Figure 5.1. When the word line goes high, a resistor divider will be formed by the pass-transistor and the pull-down transistor. As a result, the voltage of the T node will increase. In a fault free cell, the voltage will increase just some mVs, which is within the design specification of the cell. However, when the pull-down transistor resistivity is too high (due to OC3), the voltage will increase above the threshold voltage and consequently the cell will flip.

  In case the resistance value of OC3 is very high (i.e, Region VI), the cell will suffer from a data retention fault; i.e., it will not be able to maintain its state 0. Due to the broken V_{ss} path, the cell flips after a certain period T to 1 ($< 0_T/1/- >_v$). The time period T depends on the resistance of the open, as well as on the leakage current at the node.

- Region IV: In this case R_{op} is not very high, such that the flipping happens relatively slowly. As a consequence the sense amplifier read the *correct* data (i.e., the value before the cell flips); that is : $< 0r0/1/0 >_v$. Therefore, the fault will not be directly visible at the output. In order to detect this fault, another read operation has to be performed to the defective cell.

- Region III: In this case all single-port operations pass correctly; i.e., the cell does not flip due to a read operation. However, the content of the cell is well disturbed by the read operation. When two simultaneous $0r0$ operations (i.e., '$0r0 : 0r0$') are performed, the cell will flip and and the sense amplifiers outputs are set to *incorrect* data (i.e., $< 0r0 : 0r0/1/1 >_v$). When '$0r0 : 0r0$' is performed, two voltage dividers, which have an additive effect on the cell's node, will be formed. The resistance value of OC3 which was not able to flip the cell during a singe '$0r0$' operation, can in that case cause the cell to flip since two voltage dividers increase the voltage of the cell's node above the threshold.

Table 5.3. Simulation results for some SDs; d=don't care value ($d \in \{0,1\}$)

Label	R_{df} region	Fault behavior	Compl. behavior	Class	Fault model
OC3	Region I	wF	wF	-	-
	Region II	$< 0r0 : 0r0/1/0 >_v$	$< 1r1 : 1r1/0/1 >_v$	2PF1	wDRDF&wDRDF
	Region III	$< 0r0 : 0r0/1/1 >_v$	$< 1r1 : 1r1/0/0 >_v$	2PF1	wRDF&wRDF
	Region IV	$< 0r0/1/0 >_v$	$< 1r1/0/1 >_v$	1PF1	DRDF
	Region V	$< 0r0/1/1 >_v$	$< 1r1/0/0 >_v$	1PF1	RDF
	Region VI	$< 0r0/1/1 >_v$	$< 1r1/0/0 >_v$	1PF1	RDF
		$< 0_T/1/- >_v$	$< 1_T/0/- >_v$	1PF1	DRF
OC13	Region I	wF		-	-
	Region II	$< 0r0/?/? >$		1PF1	URF
		$< 1r1/?/? >$		1PF1	URF
		$< 1_T/?/- >$		1PF1	DRF
		$< 0_T/?/- >$		1PF1	DRF
BC1	Region I	$<dw0/?/- >_v$		1PF1	UWF
		$<dw1/?/- >_v$		1PF1	UWF
	Region II	wF		-	-
BC5	Region I	$\{< 0w1/0/- >_v, < 1w0/1/- >_v$			
		$< 1r1/1/? >_v, < 0r0/0/? >_v\}$		1PF1	NAF
	Region II	$< 0r0/0/? >_v$		1PF1	RRF
		$< 1r1/1/? >_v$		1PF1	RRF
	Region III	wF		-	-
BC7	Region I	$< 0r0 : 0w1/0/- >_v$		2PF1	wRDF&wTF
		$< 1r1 : 1w0/1/- >_v$		2PF1	wRDF&wTF
		$<dw1 : 1r1/0/0 >_{a,v}$		2PF2	wCFds&wRDF
		$<dw0 : 0r0/1/1 >_{a,v}$		2PF2	wCFds&wRDF
	Region II	$<dw1 : 1r1/0/0 >_{a,v}$		2PF2	wCFds&wRDF
		$<dw0 : 0r0/1/1 >_{a,v}$		2PF2	wCFds&wRDF
	Region III	$<dw1 : 1r1/1/0 >_{a,v}$		2PF2	wCFds&wIRF
		$<dw0 : 0r0/0/1 >_{a,v}$		2PF2	wCFds&wIRF
	Region IV	$<dw1 : 1r1/1/? >_{a,v}$		2PF2	wCFds&wRRF
		$<dw0 : 0r0/0/? >_{a,v}$		2PF2	wCFds&wRRF
	Region V	wF		-	-
BC8	Region I	$< 1w0/1/- >_v$	$< 0w1/0/- >_v$	1PF1	TF
		$< 1r1/0/0 >_v$	$< 0r0/1/1 >_v$	1PF1	RDF
	Region II	$< 1r1/0/0 >_v$	$< 0r0/1/1 >_v$	1PF1	RDF
	Region III	$< 1r1/1/0 >_v$	$< 0r0/0/1 >_v$	1PF1	IRF
	Region VI	$< 1r1/1/? >_v$	$< 0r0/0/? >_v$	1PF1	RRF
	Region V	wF	wF	-	-
rBCC1	Region I	$< 0;1/0/- >_{a,v}$	$< 1;0/1/- >_{a,v}$	1PF2	CFst
	Region II	$< 0;1r1/0/0 >_{a,v}$	$< 1;0r0/1/1 >_{a,v}$	1PF2	CFrd
		$< 0r0;1/0/- >_{a,v}$	$< 1r1;0/1/- >_{a,v}$	1PF2	CFds
	Region III	$< 0;1r1 : 1r1/0/0 >_{a,v}$	$< 1;0r0 : 0r0/1/1 >_{a,v}$	2PF2	wCFrd&wRDF
		$< 0r0 : 0r0;1/0/- >_{a,v}$	$< 1r1 : 1r1;0/1/- >_{a,v}$	2PF2	wCFds&wCFds
		$<dw0 : rd;1/0/- >_{a,v}$	$<dw1 : rd;0/1/- >_{a,v}$	2PF2	wCFds&wCFds
	Region IV	$< 0;1r1 : 1r1/0/1 >_{a,v}$	$< 1;0r0 : 0r0/1/0 >_{a,v}$	2PF2	wCFdrd&wDRDF
		$< 0r0 : 0r0;1/0/- >_{a,v}$	$< 1r1 : 1r1;0/1/- >_{a,v}$	2PF2	wCFds&wCFds
		$<dw0 :drd;1/0/- >_{a,v}$	$<dw1 :drd;0/1/- >_{a,v}$	2PF2	wCFds&wCFds
	Region V	wF	wF	-	-
rBCC3	Region I	$< 0/1/- >_v$	$< 1/0/- >_v$	1PF1	SF
	Region II	$< 1w0/1/- >_v$	$< 0w1/0/- >_v$	1PF1	TF
		$<dw0; 1/0/- >$	$<dw1; 0/1/- >$	1PF2	CFds
		$< 0; 1r1/1/0 >_{a,v}$	$< 1; 0r0/0/1 >_{a,v}$	1PF2	CFir
	Region III	$< 1; 1w0/1/- >_{a,v}$	$< 0; 0w1/0/- >_{a,v}$	1PF2	CFtr
		$<dw0; 1/0/- >_{a,v}$	$<dw1; 0/1/- >_{a,v}$	1PF2	CFds
		$< 0; 1r1/1/0 >_{a,v}$	$< 1; 0r0/0/1 >_{a,v}$	1PF2	CFir
	Region IV	$<dw0; 1/0/- >_{a,v}$	$<dw1; 0/1/- >_{a,v}$	1PF2	CFds
		$< 0; 1r1/1/0 >_{a,v}$	$< 1; 0r0/0/1 >_{a,v}$	1PF2	CFir
	Region V	$< 0; 1r1/1/0 >_{a,v}$	$< 1; 0r0/0/1 >_{a,v}$	1PF2	CFir
	Region VI	$< 0; 1r1/1/? >_{a,v}$	$< 1; 0r0/0/? >_{a,v}$	1PF2	CFrr
		$<dw0 :drd; 1/0/- >_{a,v}$	$<dw1 :drd; 0/1/- >_{a,v}$	2PF2	wCFds&wCFds
	Region VII	wF	wF	-	-
cBCC1	Region I	$< 0;1/0/- >_{a,v}$	$< 1;0/1/- >_{a,v}$	1PF2	CFst
	Region II	$< 0;1r1/0/0 >_{a,v}$	$< 1;0r0/1/1 >_{a,v}$	1PF2	CFrd
	Region III	$< 0;1r1/0/1 >_{a,v}$	$< 1;0r0/1/0 >_{a,v}$	1PF2	CFdrd
	Region IV	$< 0;1r1 : 1r1/0/0 >_{a,v}$	$< 1;0r0 : 0r0/1/1 >_{a,v}$	2PF2	wCFrd&wRDF
	Region V	$< 0;1r1 : 1r1/0/1 >_{a,v}$	$< 1;0r0 : 0r0/1/0 >_{a,v}$	2PF2	wCFdrd&wDRDF
	Region VI	wF	wF	-	

- Region II: In this case R_{op} is sufficiently low, such that the flipping during '0r0 : 0r0' happens relatively slow. As a consequence the sense amplifiers read the *correct* data; i.e., $< 0r0 : 0r0/1/0 >_v$.

- Region I: In this case, the cell does not flip. However, the true node is disturbed by the '0r0 : 0r0' operation; i.e., its voltage raises, but not enough to flip the cell. This can be considered as a weak fault which is not be visible at the sense amplifier outputs using either single-port or two-port operations.

5.3 Realistic fault models

The electrical faults caused by opens, shorts and bridges, and expressed in terms of FPs, can be translated into sets of *realistic* functional fault models (FFMs); whereby a FFM is defined as a non-empty set of FPs. Based on the number of ports required in order to sensitize a fault, the realistic FFMs for memory cell array faults (MCAFs) in 2P memories can be classified into single-port faults (1PFs) and two-port faults (2PFs). The 1PFs are faults that can be sensitized using single-port operations; while the 2PFs can *not* be sensitized using single-port operations: they require the use of the two ports of the memory *simultaneously*. In the following, the two realistic classes of faults will be discussed in detail; in addition, defects causing each realistic FFM will be given.

5.3.1 Realistic single-port faults

The derived 1PFs based on the found simulation results are divided into 1PFs involving a single-cell (1PF1s) and 1PFs involving two cells (1PF2s); see Figure 5.3. This classification matches the classification done in Section 3.3 for the theoretical framework of all possible 1PFs. However, not all introduced theoretical FFMs have been found through the electrical simulation. Below realistic two fault subclasses 1PF1 and 1PF2 will be discussed.

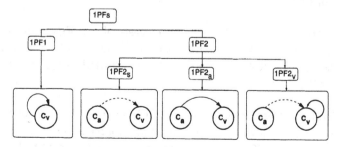

Figure 5.3. Classification of realistic 1PFs

Realistic 1PF1 faults

These are FFMs consisting of single-port, single cell FPs; they have the property
that the cell used for sensitizing the fault is the same cell as where the fault appears.
Based on the simulation results (see also the sixth column of Table 5.3 which shows
the FFM to which each FP belongs), the 1PF1s of Table 5.4 have been derived; see
also Table 3.2 which shows all possible 1PF1s.

Table 5.4. List of realistic 1PF1s; d =don't care $(d \in \{0,1\}$

#	FFM	Fault primitives
1	SF	$< 1/0/->, < 0/1/->$
2	TF	$< 0w1/0/->, < 1w0/1/->$
3	RDF	$< 0r0/1/1 >, < 1r1/0/0 >$
4	DRDF	$< 0r0/1/0 >, < 1r1/0/1 >$
5	IRF	$< 0r0/0/1 >, < 1r1/1/0 >$
6	RRF	$< 0r0/0/? >, < 1r1/1/? >$
7	UWF	$<dw0/?/->, <dw1/?/->$
8	URF	$< 0r0/?/? >, < 1r1/?/? >$
9	NAF	$\{< 0w1/0/->, < 1w0/1/->, < rx/x/? >\}$
10	DRF	$< 1_T/0/->, < 0_T/1/->, < x_T/?/->$

1. State Fault (SF): it has two FPs, and can be caused by defects like: (a) Cross
   coupling between one of the nodes of the cell and the opposite invertor is broken
   (OC6), (b) Short at one of the nodes of the cell (SC1, SC2), (c) Bridge between
   a node and a bit line of adjacent cell (rBCC3, rBCC4), etc.

2. Transition Fault (TF): it consists of two FPs, and can be caused by different
   defects like: (a) Pass transistor connection to the node of the cell broken (OC9),
   (b) Pass transistor connection to the bit line broken (OC10), (c) Short at one
   of the bit lines of the cell (SB1, SB2), (d) Bridge between a word line and a bit
   line (BC8, BC9), etc.

3. Read Destructive Fault (RDF) [5]: it consists of two FPs, and can be caused
   by different defects like: (a) Drain/source of the pull-down transistor of the
   cell broken (OC3/OC4), (b) Node of the cell shorted to V_{ss} (SC2), (c) Bit line
   shorted to V_{ss} (SB2), etc.

4. Deceptive Read Destructive Fault (DRDF) [5]: it consists of two FPs, and can
   be caused by the following defects: (a) Drain/source of the pull-down transistor
   of the cell broken (OC3/OC4), (b) Node of the cell shorted to V_{ss} (SC2), and
   (c) A bridge between a node of a cell and a word line on an adjacent cell in the
   same row(cBCC3).

5. Incorrect Read Fault (IRF): it consists of two FPs, and can be caused by the following defects: (a) Short between bit line and V_{ss} (SB2), (b) Bridge between a node of a cell at one side with its bit line at the other side (BC3), and (c) Bridge between a word line and a bit line (BC8, BC9).

6. Random Read Fault (RRF): it consists of two FPs, and can be caused by different defects like: (a) Pass transistor connection to the node of the cell broken (OC9). (b) Gate of a pass transistor broken (OC11), (c) Bit line at the read side circuit broken (OB_r), (d) Bit line shorted to V_{ss} (SB2), (e) Bridge between the two bit lines (belonging to the same port) of the cell (BC5), etc.

7. Undefined Write Fault (UWF): it consist of two FPs. Note that in Table 5.4, e.g., $< dw0/?/- >$ denotes only one FPs; this is because the initial data in the cell is *irrelevant* ($d=$ *don't care*). the write operation is used to sensitize the fault. The UWF can be caused by a bridge between the true and the false node of the cell (BC1).

8. Undefined Read Fault (URF): it consist of two FPs, and is caused by an open in the V_{ss} line of the cell (OC13), or an open in the V_{ss} line shared by physical adjacent cells (OC15).

9. No Access Fault (NAF): it consists of four FPs that occur simultaneously, and can be caused by the following defects: (a) Word line broken (OW), (b) Word line shorted to V_{ss} (SW2), and (c) Bridge between the two bit lines (belonging to the same port) of the cell (BC5).

10. Data Retention Fault (DRF): it consists of four FPs, and can be caused by: (a) Source/drain of the pull-up transistor of the cell broken (OC1, OC2), (b) Drain/source of the pull-down transistor of the cell broken (OC3, OC4), (c) Gate of the pull up transistor of the cell broken (OC5), (d) V_{cc} or V_{ss} path of the cell broken (OC12, OC13, OC14, OC15), etc.

Realistic 1PF2 faults

The realistic 1PF2 faults are also divided, depending on the cell to which the sensitizing operation is applied, into three types: $1PF2_s$, $1PF2_a$, and $1PF2_v$. The list of realistic 1PF2s, derived based on the simulation, is given in Table 5.5; see also Table 3.4 which shows all theoretical possible 1PF2s. Below the realistic 1PF2s are given together with defects causing them (see also the sixth column of Table 5.3).

Realistic $1PF2_s$: This type consists only of one FFM: State Coupling Fault (CFst), with four FPs. It can be caused by a bridge between nodes of adjacent cells in the same row (rBCC1, rBCC2), in the same column (cBCC1, cBCC2), or on the same diagonal (dBCC1, dBCC2).

Table 5.5. List of realistic 1PF2s; d=don't care value and $x \in \{0,1\}$

#	FFM	Fault primitives
1	CFst	$< 0; 0/1/- >$, $< 0; 1/0/- >$, $< 1; 0/1/- >$, $< 1; 1/0/- >$
2	CFds	$<dwx; 0/1/- >$, $<dwx; 1/0/- >$, $<xrx; 0/1/- >$, $<xrx; 1/0/- >$
3	CFtr	$< 0; 0w1/0/- >$, $< 1; 0w1/0/- >$, $< 0; 1w0/1/- >$, $< 1; 1w0/1/- >$
4	CFrd	$< 0; 0r0/1/1 >$, $< 1; 0r0/1/1 >$, $< 0; 1r1/0/0 >$, $< 1; 1r1/0/0 >$
5	CFdrd	$< 0; 0r0/1/0 >$, $< 1; 0r0/1/0 >$, $< 0; 1r1/0/1 >$, $< 1; 1r1/0/1 >$
6	CFir	$< 0; 0r0/0/1 >$, $< 1; 0r0/0/1 >$, $< 0; 1r1/1/0 >$, $< 1; 1r1/1/0 >$
7	CFrr	$< 0; 0r0/0/? >$, $< 1; 0r0/0/? >$, $< 0; 1r1/1/? >$, $< 1; 1r1/1/? >$

Realistic 1PF2$_a$: This type consists also only of one FFM: Disturb Coupling Fault (CFds), with only eight FPs. Note that in Table 5.5, e.g., $< dwx; 0/1/- >$ denotes two FPs; this is because d is irrelevant (don't care), and $x \in \{0,1\}$. Note also that $< xrx; 0/1/- >$ denotes two PFs. A CFds can be caused by different defects like: (a) Gate of the pass transistors of the cell floating (OW), (b) Word line shorted to V_{cc} (SW1), (c) Bridge between a node of the cell and its bit line (BC2, BC3), (d) Bridge between nodes of cells belonging to the same row (rBCC1, rBCC2), etc.

Realistic 1PF2$_v$: This type consists of five FFMs:

1. Transition Coupling Fault (CFtr): It consists of four FPs, and can be caused by bridges between a node of a cell and a bit line of an adjacent cell belonging to another column (rBCC3, rBCC4).

2. Read Destructive Coupling Fault (CFrd): It consists of four FPs, and can be caused by a bridge between nodes of adjacent cells in the same row (rBCC1, rBCC2), in the same column (cBCC1, cBCC2), or on the same diagonal (dBCC1, dBCC2).

3. Deceptive Read Destructive Coupling Fault (CFdrd): It consists of four FPs, and can be caused by bridges between adjacent cells belonging to the same column (cBCC1, cBCC2) or to the same diagonal (dBCC1, dBCC2).

4. Incorrect Read Coupling Fault (CFir): It consists of four FPs, and can be caused by the following defects: (a) Bridge between a node of the cell and its bit line at the same side (BC2), (b) Bridge between a node of the cell and its bit line at the opposite side (BC3), (c) Bridge between a node of a cell and a bit line of an adjacent cell belonging to another column (rBCC3, rBCC4).

5. Random Read Coupling Fault (CFrr): It consists of four FPs, and can be caused by the following defects: (a) Gate of a pass transistor broken (OC11), (b) Bridge between a node of the cell and its bit line at the same/opposite side (BC2/BC3), (c) Word line broken (OW), (d) Bridge between a node of

a cell and a bit line of an adjacent cell belonging to another column (rBCC3, rBCC4).

5.3.2   Realistic two-port faults

Two-port faults (2PFs) can not be sensitized using single-port operations; they require the use of the two ports simultaneously. In Section 3.4, all possible 2PFs have been divided into three classes; see Figure 3.3. The 2PFs involving a singe-cell (2PF1s), the 2PFs involving two cells (2PF2s) and the 2PFs involving three cells (2PF3s). In addition, the 2PF2s are divided, depending on to which cells the two simultaneous operations are applied (to the a-cell and/or to the v-cell), into three types: the $2PF2_a$, $2PF2_v$, and $2PF2_{av}$. Moreover, the introduced FFMs for each class have been considered as a combination of *two weak faults*. The fault effect of a weak fault can not fully sensitize a fault; however, the fault effects of two or more weak faults may be additive and hence fully sensitize a fault when the weak faults are sensitized simultaneously.

The realistic 2PFs, derived based on the found simulation results, are only divided into two classes (see Figure 5.4): the 2PF1s and the 2PF2s with three types (i.e., the $2PF2_a$, $2PF2_v$, and $2PF2_{av}$). Note that the 2PF3s, which were introduced theoretically, were found to be *not realistic*; this is because no FP belonging to that class is found during the simulation. In addition, not all introduced theoretical FFMs for 2PF1s and 2PF2s (see Table 3.6 and Table 3.10) are found to be realistic. The realistic 2PFs are given in Table 5.6, and will be listed below together with defects causing them [28, 26, 29]

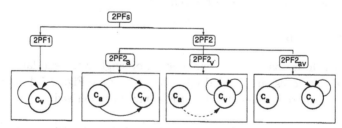

Figure 5.4. Classification or realistic 2PFs

Realistic 2PF1 faults

Based on the simulation results of the SDs, three realistic FFMs have been established (see also the sixth column of Table 5.3):

1. wRDF&wRDF: It consists of two FPs, and can be caused by the following defects: (a) Drain/source of the pull-down transistor of the cell broken

Table 5.6. List of realistic 2PFs; d =don't care value and $x \in \{0,1\}$

Class	#	FFM	Fault primitives
2PF1	1	wRDF&wRDF	$< 0r0 : 0r0/1/1 >, < 1r1 : 1r1/0/0 >$
	2	wDRDF&wDRDF	$< 0r0 : 0r0/1/0 >, < 1r1 : 1r1/0/1 >$
	3	wRDF&wTF	$< 0r0 : 0w1/0/- >, < 1r1: 1w0/1/- >$
$2PF2_a$	4	wCFds&wCFds	$< dw0 : drd; 0/1/- >, < dw0; drd; 1/0/- >, < dw1 : drd; 0/1/- >,$ $< dw1; drd; 1/0/- >, < xrx : xrx; 0/1/- >, < xrx : xrx; 1/0/- >$
$2PF2_v$	5	wCFrd&wRDF	$< 0; 0r0 : 0r0/1/1 >, < 1; 0r0 : 0r0/1/1 >,$ $< 0; 1r1 : 1r1/0/0 >, < 1; 1r1 : 1r1/0/0 >$
	6	wCFdrd&wDRDF	$< 0; 0r0 : 0r0/1/0 >, < 1; 0r0 : 0r0/1/0 >,$ $< 0; 1r1 : 1r1/0/1 >, < 1; 1r1 : 1r1/0/1 >$
$2PF2_{av}$	7	wCFds&wRDF	$< dw0 : 0r0/1/1 >, < dw1 : 0r0/1/1 >,$ $< dw0 : 1r1/0/0 >, < dw1 : 1r1/0/0 >$
	8	wCFds&wIRF	$< dw0 : 0r0/0/1 >, < dw1 : 0r0/0/1 >,$ $< dw0 : 1r1/1/0 >, < dw1 : 1r1/1/0 >$
	9	wCFds&wRRF	$< dw0 : 0r0/0/? >, < dw1 : 0r0/0/? >,$ $< dw0 : 1r1/1/0 >, < dw1 : 1r1/1/0 >$

(OC3/OC4). (b) True or false node shorted to V_{ss} (SC2), and (c) Short between a cell's node and a word line of a cell in another row (cBCC3).

2. wDRDF&wDRDF: It consists of two FPs, and can be caused by the same defects causing wRDF&wRDF but with different resistance values of the defect.

3. wRDF&wTF: It consists of two FPs, and can be caused by bridges between bit lines belonging to the same column, to different ports and to different sides; i.e., true side and false side of the cell (BC7).

Realistic 2PF2 faults

The 2PF2s, which are based on a combination of single-cell weak faults and weak faults involving two cells, are divided into three types; see Figure 5.4. Based on the FPs found by simulation, the following 2PF2s have been derived.

Realistic $2PF2_a$: This type consists only of one FFM: wCFds&wCFds, with eight FPs. Note that in Table 5.6, the '$< dw0 : drd; 0/1/- >$' denotes one FP since the read data (which is the initial stored data in the cell) is irrelevant (d=don't care value), the read operation is used to sensitize the fault. Note also that the '$< xrx : xrx; 1/0/- >$' denotes two FPs since $x \in \{0,1\}$. The $2PF2_a$ can be caused by the following defects: (a) Bridges between nodes of adjacent cells belonging to the same row (rBCC1, rBCC2), (b) Bridges between between bit lines of adjacent cells, belonging to the same or different ports (rBCC3, rBCC4, rBCC5, rBCC6, rBCC7, rBCC8).

Realistic 2PF2$_\mathbf{v}$: This type consists of two FFMs, each with four FPs; namely the wCFrd&wRDF and wCFdrd&wDRDF. Such faults are caused by bridges between physically adjacent cells belonging to the same row (rBCC1, rBCC2), to the same column (cBCC1, cBCC2), or on the same diagonal (dBCC1, dBCC2).

Realistic 2PF2$_\mathbf{av}$: This type consists of three FFMs: (a) wCFds&wRDF, (b) wCFds&wIRF, and (c) wCFds&wRRF; each with four PFs. Such faults are caused by: (a) bridges between bit lines belonging to the same column and different ports (BC6, BC7), or (b) bridges between bit lines belonging to different column and different ports (cBCC6, cBCC8).

It should be noted that the above 2PFs are valid for 2P memories, which support simultaneous reading and writing of the same location, whereby the read data is discarded; i.e., a write operation has a high priority. If this is not supported, then the FFM: wRDF&wTF will not be realistic. In addition, the FFM: wCFds&wCFds will consist only of the FPs sensitized by simultaneous read operations to the same location.

5.4 Fault probability analysis

In order to determine the importance of each FFM of Section 5.3, their probability of occurrence will be calculated using two approaches: first by assuming that all modeled SDs (i.e., opens, shorts, and bridges) are equally likely to occur, and then by using Inductive Fault Analysis (IFA).

5.4.1 SDs equally likely to occur

First, the probability of occurrence of each FP will be calculated based on the simulation results (i.e., the fault behavior and the corresponding resistance range), and by using a resistance distribution of a defect based on industrial data[2][70]. This distribution shows that the occurrence probability of a defect with a low resistance value is high compared with a defect with a high resistance value. Thereafter, the probability of each FFM is determined by summing the probabilities of all FPs (which may be caused with different SDs) constituting this FFM; remember that a FFM is a non empty set of FPs. We will focus on the relative probability of occurrence of each FFM (i.e, 1PF1 like SF and 2PF1 like wRDF&wRDF) compared with the other realistic FFMs within all memory cell array FFMs caused by all opens, shorts, and bridges.

Table 5.7 lists all FFMs that can be sensitized together with their probabilities. Since in our simulation, each simulated modeled SD is a representative of a group,

---

[2]This data is process specific and Intel proprietary

Table 5.7. Probabilities of the FFMs in 2P memories (SDs equal likely)

1PFs				2PFs			
1PF1s		1PF2s		2PF1s		2PF2s	
FFM	Prob. %	FFM	Prob. %	FFM	Prob. %	FFM	Prob. %
SF	29.921	CFst	11.805	wRDF&wRDF	1.573	wCFds&wCFds	2.234
TF	10.697	CFds	6.819	wDRDF&wDRDF	0.554	wCFrd&wRDF	0.617
RDF	5.418	CFtr	0.839	wRDF&wTF	0.371	wCFdrd&wDRDF	0.132
DRDF	0.623	CFrd	0.453			wCFds&wRDF	3.203
IRF	7.517	CFdrd	0.101			wCFds&wIRF	6.003
RRF	2.606	CFir	3.758			wCFds&wRRF	1.485
UWF	0.912	CFrr	0.681				
URF	0.031						
NAF	1.048						
DRF	0.598						
Total	59.372		24.456	Total	2.498	Total	13.674

the number of modeled defects within a group has been taken into account. Remember that a group is a set of SDs with a similar, and/or a complementary, and/or an interchanged, and/or an interchanged complementary fault behavior; e.g., OC9 is a representative of the set $\{OC9, OC9s, OC9c, OC9cs\}$; see Section 4.2 and Section 4.3. The table shows that the probability of 1PF1s (59.372%) is very high compared with 1PF2s, 2PF1s, or 2PF2s. In addition, it shows that the percentage of 2PFs, which is 16.172%, can not be ignored for tests for 2P memories. The conventional single-port (SP) tests can not detect such faults, and as a consequence the fault coverage will be negatively impacted if 2PFs are not taken into consideration.

Another aspect that shows the importance of taking 2PFs into consideration is the impact on the *Reject Ratio (RR)*. The RR is the fraction of the defective parts that pass all tests and hence is sold to customers. This ratio, given in terms of *defects per million (DPM)*, can be estimated based on the JSSC model [51]:

$$RR = \frac{(1-FC)*(1-Y)*e^{(-FC.(N_0-1))}}{Y+(1-FC)*(1-Y)*e^{(-FC.(N_0-1))}}$$

whereby FC is the fault coverage; Y is the yield and N_0 is the average number of faults per defective die.

Assume that the yield, by taking only 1PFs into consideration, is $Y_{1PFs} = 0.95$, and that the yield, by taking both 1PFs and 2PFs into consideration, is $Y_{1PFs+2PFs} = 0.95 - 0.95 * 16.172\% = 0.796$ (note that 16.172% is the percentage of 2PFs). Additionally, assume that the fault coverage, when only 1PFs are taken into consideration, is $FC_{1PFs} = 83.828\%$; this is the maximal fault coverage that can be reached with SP tests since 2PFs form 16.172% of all faults. Furthermore, when conventional SP tests as well as special tests for 2PFs will be used, then in addition to 1PFs, assume that 95% of the 2PFs will be detected. That means that the fault coverage will be $83.828\% + 0.95 * 16.172\% \simeq 99.191\%$.

Based on the the above formula and assumptions, the RR when only 1PFs are considered (RR_{1PFs}), and the RR when both 1PFs and 2PF1s are considered ($RR_{1PFs+2PFs}$) can be calculated. Table 5.8 shows RR_{1PFs} and $RR_{1PFs+2PFs}$ for different values of N_0[3]; in the table, RF denotes the reduction factor. It can be clearly seen from the comparison of the two reject ratios, that by taking the 2PFs into consideration, the RR can be reduced with a factor 7.2 on the average.

Table 5.8. RF as a function of N_0; (SDs equal likely)

N_0	RR_{1PFs}	$RR_{1PFs+2PFs}$	RF
3	1606.759	281.867	5.700
4	687.907	104.545	6.580
5	297.604	38.771	7.676
6	128.722	14.378	8.953

5.4.2   Extraction of SDs by IFA

The IFA tool *Carafe* will be used to extract SDs from the layout. Carafe uses the circuit layout, manufacturing process specifications and defect information to determine which SD can occur in the circuit layout. This tool was developed at University of California, at Santa Cruz [37]. It generates a list of opens, shorts, and bridges, with their probability of occurrence.

Two different SRAM layout designs (ML and WL), designed and processed using the same process technology, are chosen; the size of ML is 32 Kbits (i.e., 128 rows and 256 columns) and that of WL is 64 Kbits (128 rows and 512 columns). The two layouts implement the same circuit function; see Figure 5.1. To extract SDs from the layouts, Carafe has been used. Carafe extracts SDs from the layout, and orders them by *weighted critical area (WCA)*. The WCA is calculated based both on the *critical area* of the defect and on the defect statistics from the fab for the different layers and defect sizes. The critical area of a bridge (or a short) is calculated as the area where the center of a defect of a certain size could land to cause two (or more) nodes to bridge/short together. It should be noted that Carafe does *not* deal with *partial* opens (i.e., opens with $R_{op} < \infty\Omega$); it deals only with bridges, shorts, and complete opens (i.e., opens with $R_{op} = \infty\Omega$). Therefore, our analysis with IFA will be restricted to bridges and shorts only.

From the Carafe's analysis of the two layouts, the total WCA of WL was almost 7 times smaller than that of ML. This means that the defect sensitivity of ML is higher than that of WL. In addition, the IFA results show that the total WCA of multiple shorts/bridges (i.e., bridges/shorts involving more than two nodes) consist of only 2% and 5% of the total WCA of shorts and bridges (requiring two nodes) for

---

[3]these values are based on Intel experimental data

ML and WL, respectively. This explains our assumption of simulating a single SD involving at the most *two nodes* at a time.

The extracted shorts and bridges have been translated from the layout level to the electrical level. Their number can be reduced since some SDs at the layout level represent the same SDs at the electrical level. Thereafter, the WCA for which a certain FP is sensitized (denoted as WCA_{df}), in the presence of any extracted SD, is calculated based on:

a. WCA for the SD (found by Carafe)

b. the resistance range for which the FP is sensitized (obtained from the simulation results),

c. the resistance distribution of a defect (statistics obtained from the lab)

The probability of the FP is then determined by dividing WCA_{df} by the total WCA sensitizing all FPs. Finally, the probability of each FFM is calculated, by summing the occurrence probabilities of all FPs that the FFM consists of.

Table 5.9. Probabilities of FFMs based on IFA for ML and WL 2P layouts

1PFs Probability %					
1PF1s			1PF2s		
FFM	ML	WL	FFM	ML	WL
SF	29.636	10.737	CFst	0.022	1.439
TF	16.513	24.035	CFds	5.227	5.661
RDF	11.561	13.041	CFtr	0.000	0.000
DRDF	0.141	0.048	CFrd	0.002	0.089
IRF	11.264	25.336	CFdrd	0.001	0.015
RRF	1.275	1.667	CFir	5.657	3.051
UWF	9.071	5.785	CFrr	2.245	1.221
URF	n.a	n.a			
NAF	1.795	6.460			
DRF	n.a	n.a			
Total	81.256	87.109	Total	13.153	11.504

2PFs Probability %					
2PF1s			2PF2s		
wRDF&wRDF	2.835	0.969	wCFds&wCFds	0.000	0.010
wDRDF&wDRDF	0.944	0.323	wCFrd&wRDF	0.001	0.066
wRDF&wTF	0.704	0.000	wCFdrd&wDRDF	0.001	0.019
			wCFds&wRDF	0.118	0.000
			wCFds&wIRF	0.752	0.000
			wCFds&wRRF	0.236	0.000
Total	4.483	1.292	Total	1.108	0.095

Table 5.9 lists the probability of each FFM for the ML and WL layouts. Since the major URFs and DRFs are caused by partial opens, and Carafe does not deal

with such defects, their probabilities can not be determined for ML and WL; they are given in the table as 'not applicable (n.a)". The table clearly shows that the probability of occurrence of the FFMs are layout dependent. A FFM which may have a very low probability for a certain layout, can have a high probability for other layouts; for instance, the $2PF2_{av}$ consists of $0.118 + 0.752 + 0.235 = 1.106\%$ for ML, and of 0% for WL. That means, that in order to reach a very high fault coverage, a test algorithm designer has to take all realistic FFMs into consideration.

The reject ratios have been also calculated by using the found probabilities with IFA, and by calculating the yield based on an Intel methodology [85]. Table 5.10 shows RR_{1PFs} and $RR_{1PFs+2PFs}$ for different values of N_0; the table shows that by taking both 1PFs and 2PFs into consideration, the ratio can be reduced by a factor of 12.493 and 10.552, on the average, for ML, respectively WL. This is very attractive industrially.

Table 5.10. RF as a function of N_0 (based on IFA)

N_0	RR_{1PFs}		$RR_{1PFs+2PFs}$		R.F	
	ML	WL	ML	WL	ML	WL
4	400.413	81.361	34.621	7.857	11.565	10.355
5	155.812	30.351	12.808	2.894	12.165	10.487
6	60.621	11.321	4.738	1.066	12.794	10.620
7	23.580	4.223	1.753	0.393	13.451	10.745

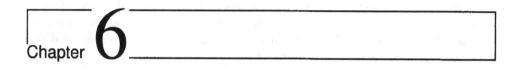

Chapter **6**

Tests for single-port and two-port SRAMs

6.1 Notation for march tests

6.2 Tests for single-port faults

6.3 Conditions for detecting two-port faults

6.4 Tests for two-port faults

6.5 Test strategy

6.6 Test results versus fault probabilities

In Chapter 5, realistic fault models have been introduced for two-port (2P) memories, based on defect injection and SPICE simulation; they are divided into single-port faults and unique faults in 2P memories. Inductive fault analysis has also been used in order to determine the importance of the introduced fault models. The results show that in order to reach a high fault coverage, new tests have to be developed for the detection of unique faults in 2P memories.

This chapter presents tests for two-port memories. It starts by introducing the notation for march tests in Section 6.1. Then, testing of single-port faults will be discussed in Section 6.3 by showing the shortcomings of the traditional tests and introducing new ones. Thereafter, the conditions to detect the unique two-port faults will be established in Section 6.3. Such conditions will be used to derive tests in Section 6.4; the classification and the comparison of these tests with traditional tests will be also addressed. Furthermore, the test strategy, which determines the methodology to be used for testing, will be discussed in Section 6.6. Finally some industrial test results will be presented in Section 6.7.

6.1   Notation for march tests

A *march test* consists of a finite sequence of march elements [87]. A *march element* is a finite sequence of operations applied to every cell in the memory before proceeding to the next cell. The way one proceeds to the next cell is determined by the address order, which can be an increasing address order (e.g., increasing address from the cell 0 to the cell $n-1$), denoted by $\Uparrow$ symbol, or a decreasing address order, denoted by $\Downarrow$ symbol, and which is the exact inverse of the $\Uparrow$ address order. When the address order is irrelevant, the symbol $\Updownarrow$ (i.e., $\Uparrow$ or $\Downarrow$) will be used. Moreover, an index can be added to the address order symbol, such that the addressing range can be specified explicitly. For example $\Uparrow_{i=0}^{n-1}$ denotes: increase the addresses from cell 0 to cell $n-1$.

An *operation* can consist of:
$w0$: write 0 into a cell.
$w1$: write 1 into a cell.
$r0$: read a cell with expected value 0.
$r1$: read a cell with expected value 1.
    In addition, the cell to which the operation is applied can be specified explicitly by subscripting the operation. For example, $w0_i$ means write 0 into cell i (i.e., c_i).

A complete march test is delimited by the '{...}' bracket pair; while a march element is delimited by the '(...)' bracket pair. The march elements are separated by semicolons, and the operations within a march element are separated by commas. The *MATS+* march test [2] $\{\Updownarrow (w0); \Uparrow (r0, w1); \Downarrow (r1, w0)\}$ consists of the march elements $\Updownarrow (w0)$, $\Uparrow (r0, w1)$, and $\Downarrow (r1, w0)$. Note that all operations of a march element are performed at a certain address, before proceeding to the next address.

The above notation, which is used for single-port (SP) memories, has to be extended, in order to be able to specify tests for multi-port (MP) memories. The extension will be done as follows:

- Ports are numbered consecutively from P_a to P_p.

- The port number to which each of the set of the parallel operations is applied is determined implicitly as follows: the first operation is applied to the first port (e.g., P_a), the second operation is applied to the second port P_b, etc. Port numbers can also be specified explicitly, by super-scripting the operation with the corresponding port number; e.g., $r0^a$ denotes that a $r0$ operation is applied to P_a.

- The operations applied in parallel to the ports are separated using colons; for example, the march element $(r0 : w1)$ denotes that a $r0$ operation is applied to the first port (e.g., P_a), and a $w1$ operation is applied to the second port P_b.

- In the case that no operation has to be applied to a particular port, the character 'n' has to be specified. For example, $(r0 : n)$ denotes that a $r0$ operation to P_a, and no operation to P_b.

- The character '$-$' denotes any allowed operation which is not in conflict with the other operation(s) of the pair. For example, $(r0 : -)$ denotes that a $r0$ operation to P_a, and any allowed operation to P_b; for example '$-$' can be replaced with '$r0$'.

- $c_{r,c}$ denotes explicitly the cell with row r and column c.

- The cell to which the operation is applied, can be specified explicitly by subscripting the corresponding operation:

    - E.g., $(r0_i : w1_j)$ denotes that a $r0$ operation is applied to a cell c_i through P_a, while a $w1$ operation is applied to a cell c_j through P_b.

    - E.g., $(r0_{r_1,c_1} : w1_{r_2,c_2})$ denotes that a $r0$ operation is applied to a cell with row r_1 and column c_1 (through P_a), while a $w1$ operation is applied to a cell with row r_2 and column c_2 (through P_b).

- The ports can be accessed with different addressing. E.g., for a two-port (2P) memory, one port can be accessed with an increasing address order, while the second with a decreasing address order. This can be specified explicitly, together with the operation(s) to be done. For example $(\Uparrow (r0) :\Downarrow (r0))$ denotes: read the cells via the two ports, using opposite address orders.

- $\Uparrow_{v=0}^{n-2}\Uparrow_{a=v+1}^{n-1}$: denotes a nested addressing sequence, whereby the cell v goes from 0 to $n-2$; and for each value of v, cell a goes from $v+1$ to $n-1$.

The following examples will clarify the introduced notation:

Example 1: $\{ \Updownarrow (w0 : n); \Uparrow (r0 : r0, w1 : n); \Downarrow (r1 : r1, w0 : n)\}$

This is the same test as MATS+, but extended for 2P memories, whereby the write operations are applied through P_a and the read operations are changed with two simultaneous read operations applied through the two ports to a single cell.

Example 2: $\{ \Updownarrow (n : w0); \Uparrow_{i=0}^{n-1} (\Uparrow_{j=i+1}^{n-1} (n : r0_j, w1_i : w1_j, w0_i : w0_j))\}$

This march test consists of two march elements. The first march element specifies a single write operation applied to all cells through P_b; while the second march element specifies a single read operation to c_j via P_b followed by simultaneous up and down transitions in all pairs of cells (c_i, c_j).

6.2   Tests for single-port faults

In Chapter 5, realistic functional fault models (FFMs) for 2P memories have been established based on defect injection and circuit simulation. Such faults have been divided into single-port faults (1PFs) and two-port faults (2PFs). The 1PFs are faults that occur in conventional SP memories; while the 2PFs are special faults for 2P memories. The 2PFs can not be sensitized using SP operations; they require the use of the two ports of the memory simultaneously. In the remainder of this section, first the effectiveness of the traditioanl memory tests regarding the realistic 1PFs of Chapter 5 will be discussed. This will show the shortcoming of such tests; therefore a new test targeting the realistic 1PFs (found based on circuit simulation) will be introduced; as well as a test detecting all possible 1PFs discussed in Chapter 3 including the faults not foud to be realsitic for the simulated design in Chapter 5. A New test targeting these

6.2.1   Effectiveness of traditional tests

Many memory test algorithms were developed to cover FFMs; most of which had a theoretical origin. The traditional ad-hoc tests, such as Walking 1/0, Galpat, Butterfly and Scan, have been used in the past to screen the faulty devices. However, the time complexity of the first two tests is completely unacceptable currently; while the fault coverage of the last two tests is not acceptable industrially.

   March tests have been introduced to detect TFs, *Inversion Coupling Faults (CFin's)*, *Idempotent Coupling Faults CFid (CFid's)*, as well as SFs. A CFin is defined as: an up (or down) transition write operation in the a-cell causes an inversion in the v-cell; i.e., the v-cell flips to 0 if its content was 1, and flips to 1 if its content was 0. A CFid is defined as: an up (or down) transition write operation in the a-cell forces a certain fixed value, 0 or 1, in the v-cell. The CFin, which has a theoretical origin, has never been shown to exist in real designs; while the FPs of the CFid are a subset of the CFds; see Section 3.3.4.

   Table 6.1 and Table 6.2 summarize the fault coverage of the most known memory tests for all possible 1PFs (see Table 3.2 and Table 3.4); only FFMs with deterministic data output at the sense amplifier are included in the tables. The FFMs with random data outputs may probabilistically detected; a test with more read operations has higher detection probability. The test length and the time complexity (T.C.) of each test is also included, where n denotes the size of the memory, R denotes the number of rows and C denotes the number of columns. The included tests are:

1. SCAN [2]: $\{\Uparrow (w0); \Uparrow (r0); \Uparrow (w1); \Uparrow (r0)\}$

2. MATS+ [59]: $\{\Updownarrow (w0); \Uparrow (r0, w1); \Downarrow (r1, w0)\}$

3. MATS++ [11]: $\{\Updownarrow (w0); \Uparrow (r0, w1); \Downarrow (r1, w0, r0)\}$

4. March C- [49, 88]: $\{\Updownarrow (w0); \Uparrow (r0, w1); \Uparrow (r1, w0); \Downarrow (r0, w1); \Downarrow (r1, w0);$
    $\Updownarrow (r0)\}$

5. PMOVI [39]: $\{\Downarrow (w0); \Uparrow (r0, w1, r1); \Uparrow (r1, w0, r0); \Downarrow (r0, w1, r1); \Downarrow (r1, w0, r0)\}$
6. March G [87]: $\{\Updownarrow (w0); \Uparrow (r0, w1, r1, w0, r0, w1); \Uparrow (r1, w0, w1);$
   $\Downarrow (r1, w0, w1, w0); \Downarrow (r0, w1, w0); \Uparrow (r0, w1, r1); \Uparrow (r1, w0, r0)\}$
7. Butterfly [88]: $\{\Uparrow (w0); \Uparrow (w1_b, \diamond(r0), r1_b, w0_b);$
   $\Uparrow (w1); \Uparrow (w0_b, \diamond(r1), r0_b, w1_b)\}$

The e.g., $w1_b$ means that a $w1$ operation is applied to the *base cell*; while $\diamond(r0)$ means that the $r0$ operations are performed to the North, East, South and West (N, E, S, W) neighbors of the *base cell*.

8. Galpat [11]: $\{\Uparrow (w0); \Uparrow_b (w1_b, \Uparrow_{-b} (r0, r1_b), w0_b);$
   $\Uparrow (w1); \Uparrow_b (w0_b, \Uparrow_{-b} (r1, r0_b), w1_b)\}$

A GALPAT has a time complexity of $O(n^2)$; it stars with initializing the memory cell array to a known data-background (e.g., $\Uparrow (w0)$). During the test, the base cell, which walks through the memory cell array, is written the data complement value (i.e., $w1_b$). Thereafter all other cells (i.e., *target cells*) are read, but after each target cell also the base cell is read (i.e., $\Uparrow_{-b} (r0, r1_b)$). The base cell is finally written to the initial data-background (e.g., $w0_b$) This whole sequence is then repeated with inverted data.

9. Walking 1/0 [88]: $\{\Uparrow (w0); \Uparrow_b (w1_b, \Uparrow (r0), r1_b, w0_b);$
   $\Uparrow (w1); \Uparrow_b (w0_b, \Uparrow (r1), r1_b, w0_b)\}$

Walking 1/0 is similar to Galpat; the difference is in reading the base cell. With Walking 1/0, after each step, all cells are read with the base cell last.

Table 6.1. Single-cell fault coverage for traditional tests

#	Tests	Test Length	T.C.	SF	TF	WDF	RDF	DRDF	IRF	FC
1	SCAN	4n	$O(n)$	2/2	1/2	0/2	2/2	0/2	2/2	7/12
2	MATS+	5n	$O(n)$	2/2	1/2	0/2	2/2	0/2	2/2	7/12
3	MATS++	6n	$O(n)$	2/2	2/2	0/2	2/2	0/2	2/2	8/12
4	March C-	10n	$O(n)$	2/2	2/2	0/2	2/2	0/2	2/2	8/12
5	PMOVI	13n	$O(n)$	2/2	2/2	0/2	2/2	2/2	2/2	10/12
6	March G	23n	$O(n)$	2/2	2/2	0/2	2/2	1/2	2/2	9/12
7	Butterfly	16n	$O(n)$	2/2	2/2	0/2	2/2	0/2	2/2	8/12
8	Galpat	6n+4nRC	$O(n^2)$	2/2	2/2	0/2	2/2	2/2	2/2	10/12
9	Walking 1/0	8n+2nRC	$O(n^2)$	2/2	2/2	0/2	2/2	0/2	2/2	8/12

In Table 6.1 and Table 6.2, "a/b" denotes that the test detects 'a' of the 'b' FPs of the corresponding FFM. E.g., March C- detects both FPs of TF, while MATS+ detects just one of them. For two-cell FFMs, each FP is divided into two sub-FPs: (a): the v-cell has a lower address than the a-cell, and (b) the v-cell has a higher address than the a-cell. For example, CFst consists of 4 FPs (see Table 3.4); considering the position of the a-cell against the v-cell leads to 8 sub-FPs. E.g., MATS+ detect 6 of 8 CFst sub-FPs. The CFds is divided into three types: CFds$_{xrx}$ whereby the fault

Table 6.2. Two-cell fault coverage for traditional tests

#	Tests	CFst	CFds			CFtr	CFwd	CFrd	CFdr	CFir	FC
			xrx	$xw\overline{x}$	xwx						
1	SCAN	5/8	2/8	1/8	0/8	2/8	0/8	4/8	0/8	4/8	18/72
2	MATS+	6/8	3/8	3/8	0/8	2/8	0/8	4/8	0/8	4/8	22/72
3	MATS++	6/8	3/8	3/8	0/8	4/8	0/8	4/8	0/8	4/8	24/72
4	March C-	8/8	8/8	8/8	0/8	8/8	0/8	8/8	0/8	8/8	48/72
5	PMOVI	8/8	8/8	7/8	0/8	8/8	0/8	8/8	6/8	8/8	53/72
6	March G	8/8	7/8	8/8	0/8	7/8	0/8	8/8	2/8	8/8	48/72
7	Butterfly	8/8	4/8	4/8	0/8	4/8	0/8	8/8	0/8	8/8	36/72
8	Galpat	8/8	8/8	4/8	0/8	4/8	0/8	8/8	4/8	8/8	44/72
9	Walking 1/0	8/8	8/8	4/8	0/8	4/8	0/8	8/8	0/8	8/8	40/72

is caused by a read operation; $\text{CFds}_{xw\overline{x}}$ whereby the fault is caused by a transition write operation; and CFds_{xwx} whereby the fault is caused by a non-transition write operation ($x \in \{0, 1\}$). The CFin, which has a theoretical origin, has never been shown to exist in real designs and therefore not included in the table; while the CFid describes the same faults as $\text{CFds}_{xw\overline{x}}$. The last column in each table (i.e., 'FC') gives the total detected FPs for the corresponding test. E.g., PMOVI detects 10/12 of single-cell and 53/72 of two-cell faults.

It is clear from the table that none of the existing tests has the capability to detect all 1PFs faults, neither the faults found to be realistic for the specific design considered for defect injection and SPICE simulation in Chapter 5. Even with using all the tests a 100% fault coverage can not be achieved; e.g., none of the tests can detect CFds_{xwx} and CFdr. This proves the need for new tests, which are given next.

6.2.2 March SR+ and March SRD+

This section give a test which targets the 1PFs found to be realistic in Chapter 5; and therefore the test is optimized for the considered design in the simulation. The test is referred to as *March SR+* (march test for simple[1] realistic faults) will be introduced; the '+' is added since that test is based on March SR [24].

$$\{ \Downarrow (w0) \; ; \; \Uparrow (r0, r0, w1, r1, r1, w0, r0) \; ; \; \Downarrow (r0)$$
$$\qquad M_0 \qquad\qquad\qquad M_1 \qquad\qquad\qquad M_2$$
$$\Uparrow (w1) \; ; \; \Downarrow (r1, r1, w0, r0, r0, w1, r1) \; ; \; \Uparrow (r1) \; \}$$
$$\qquad M_3 \qquad\qquad\qquad M_4 \qquad\qquad\qquad M_5$$

Figure 6.1. March SR+

The test is shown in Figure 6.1; it has a test length of $18n$. It detects all FFMs with a deterministic data output at the sense amplifier, except DRFs; see Table 5.4

---

[1] not linked

and Table 5.5. In addition, it may probabilistically detect the FFMs with random data outputs since each cell is read six times with an expected value '0', and six times with an expected value '1'. Note that the symmetrical structure of the test make it better suitable for BIST implementation [1].

Fault coverage of single-cell faults (1PF1s)

- All SFs, RDFs, and IRFs are detected since from each cell a 0 and a 1 is read.

- All TFs are detected because each cell is read after an up and a down transition operation; this is the case by M_1 (also by M_4).

- All DRDFs are detected because two *successive* read operations are applied to each cell (see M_1 and M_4); the first operation will sensitize the fault and the second will detect it.

- The detection of RRFs, UWFs, URFs, and NAFs can not be guaranteed. However, the test may probabilistically detect these faults since each cell is read with different data values within a single march element.

Fault coverage of two-cell faults (1PF2s)

Let's $M_{i,j}$ denotes the jth operation of march element M_i; e.g., $M_{1,3}$ denotes the third operation (i.e., '$w1$') of M_1.

- All CFst's are detected.
  - The FP '$< 0; 0/1/- >$' is detected by $M_{1,1}$ (also by $M_{1,7}$).
  - The FP '$< 0; 1/0/- >$' is detected by $M_{1,4}$.
  - The FP '$< 1; 0/1/- >$' is detected by $M_{4,4}$.
  - The FP '$< 1; 1/0/- >$' is detected by $M_{4,1}$ (also by $M_{4,7}$).

- All CFds's are detected.
  - All FPs with '1' as the value of the fault effect are sensitized by M_1; e.g., $M_{1,1}$ sensitizes '$< 0r0; 0/1/1 >$', $M_{1,3}$ sensitizes '$< w1; 0/1/1 >$', etc. All these faults will be detected by $M_{1,1}$ if the address of the v-cell is higher than the address of the a-cell (i.e., $v > a$). If $v < a$, then these faults will be detected by M_2.
  - All FPs with '0' as the value of the fault effect are sensitized by M_4; e.g., $M_{4,1}$ sensitizes '$< 1r1; 1/0/1 >$', $M_{4,3}$ sensitizes '$< w0; 1/0/1 >$', etc. All these faults will be detected by $M_{4,1}$ if $v < a$, and by M_5 if $v > a$.

- All CFtr's are detected.
  - The FP '$< 0; 0w1/0/- >$' is sensitized by $M_{1,3}$ and detected by $M_{1,4}$
  - The FP '$< 0; 1w0/1/- >$' is sensitized by $M_{1,6}$ and detected by $M_{1,7}$

    – The FP '$< 1; 0w1/0/- >$' is sensitized by $M_{4,3}$ and detected by $M_{4,4}$
    – The FP '$< 1; 1w0/1/- >$' is sensitized by $M_{4,6}$ and detected by $M_{4,7}$

- All CFrd's are detected.
  - The FP '$< 0; 0r0/1/1 >$' is detected by $M_{1,1}$ (also by $M_{1,7}$).
  - The FP '$< 0; 1r1/0/0 >$' is detected by $M_{1,4}$.
  - The FP '$< 1; 0r0/1/1 >$' is detected by $M_{4,1}$ (also by $M_{4,7}$).
  - The FP '$< 1; 1r1/0/0 >$' is detected by $M_{4,4}$.

- All CFdrd's are detected.
  - The FP '$< 0; 0r0/1/0 >$' is sensitized by $M_{1,1}$ and detected by $M_{1,2}$.
  - The FP '$< 0; 1r1/0/1 >$' is sensitized by $M_{1,4}$ and detected by $M_{1,5}$.
  - The FP '$< 1; 0r0/1/0 >$' is sensitized by $M_{4,1}$ and detected by $M_{4,2}$.
  - The FP '$< 1; 1r1/0/1 >$' is sensitized by $M_{4,4}$ and detected by $M_{4,5}$.

- All CFir's are detected.
  - The FP '$< 0; 0r0/0/1 >$' is detected by $M_{1,1}$ (also by $M_{1,7}$).
  - The FP '$< 0; 1r1/1/0 >$' is detected by $M_{1,4}$.
  - The FP '$< 1; 0r0/0/1 >$' is detected by $M_{4,1}$ (also by $M_{4,7}$).
  - The FP '$< 1; 1r1/1/0 >$' is detected by $M_{4,4}$.

- The detection of CFrr can not be guaranteed. However, the test may probabilistically detect this fault since each cell is read with different data values within a single march element.

The detection of DRFs requires bringing the cell in a certain state, adding a certain delay, and thereafter reading the cell (this has to be done for both logic states of the cell). March SR+ can be extended in order to additionally detect DRFs. The result is shown in Figure 6.2, and referred as *March SRD+*. The two inserted delay elements guarantees the detection of the two FPs of the DRF with a deterministic data output at the sense amplifier. In addition, it may probabilistically detect the other two FPs with random data outputs; see also Table 5.4.

It is further important to note that March SR+ does not detect the WDF, CFwd and CFds$_{xwx}$; these faults are not found to be realistic for the considered design and therefore not taken into consideration for March SR+; the test also detecting WDF, CFwd and CFds$_{xwx}$ is given next.

$$\{ \Downarrow (w0) \;\; ; \;\; \Uparrow (r0, r0, w1, r1, r1, w0, r0) \;\; ; Del; \;\; \Downarrow (r0)$$
$$M_0 \qquad\qquad\qquad M_1 \qquad\qquad\qquad\qquad M_2$$
$$\Downarrow (w1) \;\; ; \;\; \Downarrow (r1, r1, w0, r0, r0, w1, r1) \;\; ; Del; \;\; \Uparrow (r1) \}$$
$$M_3 \qquad\qquad\qquad M_4 \qquad\qquad\qquad\qquad M_5$$

Figure 6.2. March SRD+

6.2.3 March SS

This section give a test detecting all possible 1PF1s ad described in Chapter 3. The test is of interest when some extra test time can be tolerated for covering *all* possible single-port faults. The test is given in Figure 6.3; it is called *March SS* (Static Simple; see also Figure 3.1) and has a test length of $22n$ [30]. Minimization of the test length was considered a high priority. However, M_5 can be extended (e.g., $\updownarrow (r0, r0, w0, r0, w1)$) if a regular structure is required for BIST applications.

Let $M_{i,j}$ denote the jth operation of march element M_i; e.g., $M_{1,3}$ denotes the third operation (i.e., w0) of M_1.

$$
\begin{array}{l}
\{ \updownarrow (w0) \ ; \\
\quad M_0 \\
\Uparrow (r0, r0, w0, r0, w1) \ ; \quad \Uparrow (r1, r1, w1, r1, w0) \ ; \\
\quad M_1 \qquad\qquad\qquad\qquad M_2 \\
\Downarrow (r0, r0, w0, r0, w1) \ ; \quad \Downarrow (r1, r1, w1, r1, w0) \ ; \\
\quad M_3 \qquad\qquad\qquad\qquad M_4 \\
\updownarrow (r0) \ \} \\
\quad M_5
\end{array}
$$

Figure 6.3. March SS

Fault coverage of single-cell faults (1PF1s)

March SS detects all 1PFs with a deterministic data output (see also Table 3.2):

- All SFs, RDFs and IRFs are detected since from each cell a 0 and a 1 is read.

- All TFs are detected because each cell is read after an up and a down transition write operation. The $< 0w1/0/- >$ is sensitized by $M_{1,5}$ (also by $M_{3,5}$) and detected by $M_{2,1}$ ($M_{4,1}$); while the $< 1w0/0/- >$ is sensitized by $M_{2,5}$ (also by $M_{4,5}$) and detected by $M_{3,1}$ (M_5).

- All WDFs are detected since each cell is read after a non-transition write operation; this is done by M_1 and M_2 (also by M_3 and M_4).

- All DRDFs are detected because two *successive* read operations are applied to each cell; the first read operation sensitizes the fault while the second detects it.

Fault coverage of two-cell faults (1PF2s)

March SS detects all 1PF2s with deterministic data outputs (see also Table 3.4):

- The detection of CFst's requires that four states of any two cells can be generated, and verified by a read operation [21]. The reader can easily verify by using a state diagram for March SS that all states of any two cells c_i and c_j (i.e., 00, 01, 11, 10) are generated and verified.

Table 6.3. Fault coverage of March SS

FFM	FP	$v > a$	$v < a$
CFds	$< 0r0; 0/1/- >$	$M_{1,1}/M_{1,1}$	$M_{3,1}/M_{3,1}$
	$< 0r0; 1/0/- >$	$M_{3,1}/M_{4,1}$	$M_{1,1}/M_{2,1}$
	$< 1r1; 0/1/- >$	$M_{4,1}/M_5$	$M_{2,1}/M_{3,1}$
	$< 1r1; 1/0/- >$	$M_{2,1}/M_{2,1}$	$M_{4,1}/M_{4,1}$
	$< 0w1; 0/1/- >$	$M_{1,5}/M_{1,1}$	$M_{3,5}/M_{3,1}$
	$< 0w1; 1/0/- >$	$M_{3,5}/M_{4,1}$	$M_{1,5}/M_{2,1}$
	$< 1w0; 0/1/- >$	$M_{4,5}/M_5$	$M_{2,5}/M_{3,1}$
	$< 1w0; 1/0/- >$	$M_{2,5}/M_{2,1}$	$M_{4,5}/M_{4,1}$
	$< 0w0; 0/1/- >$	$M_{1,3}/M_{1,1}$	$M_{3,3}/M_{3,1}$
	$< 0w0; 1/0/- >$	$M_{3,3}/M_{4,1}$	$M_{1,3}/M_{2,1}$
	$< 1w1; 0/1/- >$	$M_{4,3}/M_5$	$M_{2,3}/M_{3,1}$
	$< 1w1; 1/0/- >$	$M_{2,3}/M_{2,1}$	$M_{4,3}/M_{4,1}$
CFwd	$< 0; 0w0/1/- >$	$M_{3,3}/M_{3,4}$	$M_{1,3}/M_{1,4}$
	$< 1; 0w0/1/- >$	$M_{1,3}/M_{1,4}$	$M_{3,3}/M_{3,4}$
	$< 0; 1w1/1/- >$	$M_{2,3}/M_{2,4}$	$M_{4,3}/M_{4,4}$
	$< 1; 1w1/1/- >$	$M_{4,3}/M_{4,4}$	$M_{2,3}/M_{2,4}$
CFdrd	$< 0; 0r0/1/0 >$	$M_{3,1}/M_{3,2}$	$M_{1,1}/M_{1,2}$
	$< 1; 0r0/1/0 >$	$M_{1,1}/M_{1,2}$	$M_{3,1}/M_{3,2}$
	$< 0; 1r1/0/1 >$	$M_{2,1}/M_{2,2}$	$M_{4,1}/M_{4,2}$
	$< 1; 1r1/0/1 >$	$M_{4,1}/M_{4,2}$	$M_{2,1}/M_{2,2}$
CFtr	$< 0; 0w1/0/- >$	$M_{3,5}/M_{4,1}$	$M_{1,5}/M_{2,1}$
	$< 1; 0w1/0/- >$	$M_{1,5}/M_{2,1}$	$M_{3,5}/M_{4,1}$
	$< 0; 1w0/1/- >$	$M_{2,5}/M_{3,1}$	$M_{4,5}/M_5$
	$< 1; 1w0/1/- >$	$M_{4,5}/M_5$	$M_{2,5}/M_{3,1}$

- All CFds's are detected; this includes CFds's based on read operations, on transition write operations and on non-transition write operations. The first block of Table 6.3 shows by which march element (i.e., M_0 through M_5) of March SS, each FP belonging to each FFM is sensitized and detected. In the table, two cases have been distinguished: a) the v-cell has a higher address than the a-cell (i.e., $v > a$), and b) the v-cell has a lower address than the a-cell ($v < a$). In addition, in each entry the notation Sensitization/Detection is used. E.g., the $< 0r0; 0/1/- >$ is sensitized and detected by $M_{1,1}$ when $v > a$; while $< 0r0; 1/0/- >$ is sensitized by $M_{3,1}$ and detected by $M_{4,1}$ when $v > a$.

- All CFwd's are detected. The detection of CFwd's requires that each pair of cells undergoes the four states (00, 01, 10, 11), the application of a non-transition operation and thereafter a read operation. The second block of Table 6.3 shows by which march element each FP of CFwd is sensitized and detected.

- All CFdrd's, CFrd's, CFir's are detected. The detection of CFrd's and CFir's require that each pair of cells undergoes the four states (00, 01, 10, 11), and a read operation has to be performed to each of the two cells; while the detection of CFdr's require, in addition, the application of another read operation. Therefore, any test detecting CFdr also detects CFrd and CFir. The third block of Table 6.3 shows by which march element each FP of CFdr is sensitized and detected.

- All CFtr's are detected. The detection of CFtr's requires that each pair of cells undergoes the four states (00, 01, 10, 11), the application of a transition write operation to sensitize the fault, and thereafter a read operation to detect it. The fourth block of Table 6.3 shows by which march element each FP of CFtr is sensitized and detected.

Furthermore, the test may probabilistically detect the other single-cell and two-cell faults with random data outputs; see also Table 5.4. E.g., since the test consist of 7 read 0 operations, the probability to detect the RRF$=< 0r0/0/? >$ is $\frac{1}{2} + \frac{1}{4} + \frac{1}{8} + \frac{1}{16} + \frac{1}{32} + \frac{1}{64} = 0.992$

6.3 Conditions for detecting two-port faults

The two-port faults (2PFs) have been divided into 2PFs involving a single cell (2PF1s) and 2PFs involving two cells (2PF2s); see Figure 5.4. Based on to which the two simultaneous operations have to be applied in order to sensitize the fault (both to the a-cell, both to the v-cell, or one operation to the a-cell and one to the v-cell), the 2PF2s have been further divided into three types: $2PF2_a$, $2PF2_v$, $2PF2_{av}$, respectively. The established realistic FFMs for 2PFs are regiven in Table 6.4, together with their FPs. In the following, the conditions for detecting these faults will be discussed; first for 2PF1s and thereafter for 2PF2s. These conditions will be used to derive functional tests [26, 29].

6.3.1 Conditions for detecting 2PF1s

The 2PF1 fault class has the property that the fault is sensitized in the v-cell by applying two simultaneous operations to the same cell. It consists of three FFMs: wRDF&wRDF,wDRDF&wDRDF, and wRDF&wTF; see Table 6.4.

Conditions for detecting wRDF&wRDF

The wRDF&wRDF consists of two FPs; see Table 6.4. It has to be considered only for 2P memories which allow for two simultaneous read operations of a same location (i.e., address). It is detectable if the following condition is satisfied:

Table 6.4. List of 2PFs; $x \in \{0, 1\}$ and $d =$ don't care

Class	FFM	Fault primitives
2PF1	wRDF&wRDF	$< 0r0 : 0r0/1/1 >$, $< 1r1 : 1r1/0/0 >$
	wDRDF&wDRDF	$< 0r0 : 0r0/1/0 >$, $< 1r1 : 1r1/0/1 >$
	wRDF&wTF	$< 0r0 : 0w1/0/- >$, $< 1r1 : 1w0/1/- >$
$2PF2_a$	wCFds&wCFds	$< dw0 : drd; 0/1/- >$, $< dw0 : drd; 1/0/- >$
		$< dw1 : drd; 0/1/- >$, $< dw1 : drd; 1/0/- >$
		$< xrx : xrx; 0/1/- >$, $< xrx : xrx; 1/0/- >$
$2PF2_v$	wCFrd&wRDF	$< 0; 0r0 : 0r0/1/1 >$, $< 0; 1r1 : 1r1/0/0 >$,
		$< 1; 0r0 : 0r0/1/1 >$, $< 1; 1r1 : 1r1/0/0 >$
	wCFdrd&wDRDF	$< 0; 0r0 : 0r0/1/0 >$, $< 0; 1r1 : 1r1/0/1 >$,
		$< 1; 0r0 : 0r0/1/0 >$, $< 1; 1r1 : 1r1/0/1 >$,
$2PF2_{av}$	wCFds&wRDF	$< dw0 : 0r0/1/1 >$, $< dw0 : 1r1/0/0 >$,
		$< dw1 : 0r0/1/1 >$, $< dw1 : 1r1/0/0 >$
	wCFds&wIRF	$< dw0 : 0r0/0/1 >$, $< dw0 : 1r1/1/0 >$,
		$< dw1 : 0r0/0/1 >$, $< dw1 : 1r1/1/0 >$
	wCFds&wRRF	$< dw0 : 0r0/0/? >$, $< dw0 : 1r1/1/? >$,
		$< dw1 : 0r0/0/? >$, $< dw1 : 1r1/1/? >$

Condition wRDF&wRDF: Any wRDF&wRDF fault is detectable by a march test if the test contains the march element of Case A and the march element of Case B; these two march elements can be combined into a single march element.

- Case A (to detect $< 0r0 : 0r0/1/1 >_v$): $\updownarrow (..., r0 : r0, ...)$
- Case B (to detect $< 1r1 : 1r1/0/0 >_v$): $\updownarrow (..., r1 : r1, ...)$

The pair of simultaneous read operations through the two ports in each march element sensitize and detect the fault. This is because the read operations will return the non expected values.

Conditions for detecting wDRDF&wDRDF

The wDRDF&wDRDF consists of two FPs; see Table 6.4. It has to be considered only for 2P memories which allow for two simultaneous read operations of a same location. It is detectable if the following condition is satisfied:

Condition wDRDF&wDRDF: Any wDRDF&wDRDF fault will be detected by a march test which contains the two march elements of Case A and the two march elements of Case B; these four march elements may be combined into three, two, or a single march element.

- Case A (to detect $< 0r0 : 0r0/1/0 >_v$): $\updownarrow (..., r0 : r0); \updownarrow (r0 : -, ...)$.
- Case B (to detect $< 1r1 : 1r1/0/1 >_v$): $\updownarrow (..., r1 : r1); \updownarrow (r1 : -, ...)$.

The first pair of simultaneous read operations through the two ports in each first march element sensitizes the fault, which will be detected by the single read operation of each second march element. Note that the single read operation can be replaced by simultaneous read operations (i.e., '−', which denotes any allowed (non-conflicting) operation, can be replaced with a read operation). Note that any test detecting wDRDF&wDRDF will also detect wRDF&wRDF, since Condition wRDF&wRDF is a subset of Condition wDRDF&wDRDF.

Conditions for detecting wRDF&wTF

The wRDF&wTF consists of two FPs; see Table 6.4. It only applies to 2P memories which allow for simultaneous read and write of the same location, whereby the read data will be discarded. Such faults are detectable if the following condition is satisfied:

Condition wRDF&wTF: Any wRDF&wTF fault will be detected by a march test which contains the pair of march elements of Case A and of Case B. The individual march elements may be combined into one, two, or three march elements.

- Case A (to detect $< 0r0 : 0w1/0/- >_v$): $\updownarrow (..., w1 : r0); \updownarrow (r1 : -, ...)$

- Case B (to detect $< 1r1 : 1w0/1/- >_v$): $\updownarrow (..., w0 : r1); \updownarrow (r0 : -, ...)$

The first pair of simultaneous operations through the two ports in each march element sensitize the fault, which will be detected by the second march elements consisting of single read operations. Note that the single read operations can be replaced by two simultaneous read operations. Note also that in; e.g., '$w1 : r0$', the '$w1$' indicates a up-transition write operation since the initial data in the cell is 0, which is the expected read value for '$r0$'.

Conditions for detecting all 2PF1s

The above three conditions can be merged into a single condition; the result, referred to as Condition 2PF1, is shown below. A test satisfying this condition detects all 2PF1 faults.

Condition 2PF1: Any 2PF1 fault will be detected by a march test which contains both pairs of march elements of Case A (i.e., A.1 and A.2), or both pairs of march elements of Case B (i.e., B.1 and B.2). The six pairs of march elements A.1 and A.2 can be combined into one, two, three, four, or five march elements; while B.1 and B.2 can be combined into one, two, three, four, five, six, or seven elements.

- Case A:

    1. $\updownarrow (..., w0 : r1); \updownarrow (r0 : r0); \updownarrow (r0 : -, ...)$

2. $\Updownarrow (..., w1:r0); \Updownarrow (r1:r1); \Updownarrow (r1:-,...)$

- Case B:

    1. $\Updownarrow (..., r0:r0); \Updownarrow (r0:-,...); \Updownarrow (..., w1:r0); \Updownarrow (r1:-,...)$
    2. $\Updownarrow (..., r1:r1); \Updownarrow (r1:-,...); \Updownarrow (..., w0:r1); \Updownarrow (r0:-,...)$

It should be noted that the Condition 2PF1 applies to a 2P memory supporting simultaneous read and write of the same location. If this is not the case, then the FFM wRDF&wTF is not realistic; and as a consequence Condition 2PF1 can be simplified to Condition wDRDF&wDRDF.

6.3.2   Conditions for detecting 2PF2s

Depending on to which cells (to the a-cell and/or to the v-cell) the two simultaneous operations are applied, the 2PF2 subclass is divided into three types: $2PF2_a$, $2PF2_v$, and $2PF2_{av}$.

Conditions for detecting $2PF2_a$s

The $2PF2_a$ has the property that the fault is sensitized in the v-cell by applying two simultaneous operations to the a-cell. It consists of one FFM (i.e., wCFds&wCFds) with eight FPs; see Table 6.4. Remember that in the table, the '$< dw0 : drd; 0/1/->$' denotes only one FP since the read value is irrelevant (i.e., d is the don't care value); while the '$< xrx : xrx; 1/0/->$' denotes two FPs since $x \in \{0,1\}$. In order to detect any $2PF2_a$, we have:

1. to apply sensitizing operations to the a-cell c_a; $a \in \{0,1,2,...,n-2,n-1\}$.
2. to detect the fault in v-cell c_v, whereby $v \neq a$, by reading the same cell.

The order in which c_a has to be selected is not relevant. Therefore, the $\Updownarrow_{a=0}^{n-1}$ address order can be specified.

Condition 2PF2$_a$: Any $2PF2_a$ is detectable by a march test which contains both march elements of Case A and both march elements of Case B.

- Case A:

    - $\Updownarrow_{a=0}^{n-1} (r0:r0,...,w1:rd,...,r1:r1,...,w0:rd); \Updownarrow_{a=0}^{n-1} (r0:-,...)$

- Case B:

    - $\Updownarrow_{a=0}^{n-1} (r1:r1,...,w0:rd,...,r0:r0,...,w1:rd); \Updownarrow_{a=0}^{n-1} (r1:-,...)$

Condition $2PF2_a$ can be explained as follows: the operations in the first march element of Case A will sensitize all $2PF2_a$ faults when the value of the fault effect is 1, because that march element contains all sensitizing operations. If the address order of a is increasing , then Case A will detect the fault by the '$r0 : r0$' operation of the first march element if the v-cell has a higher address than the a-cell (i.e., $v > a$), and by '$r0 : -$' operation of the second march element if $v < a$. If the address order of a is decreasing, then Case A will detect the fault by the '$r0 : r0$' operation of the first march element if $v < a$, and by '$r0 : -$' operation of the second march element if $v > a$. A similar explanation can be given for Case B, which is required to sensitize and detect $2PF2_a$ faults when the value of the fault effect is 0. Note that the single read operations; e.g., the '$r0 : -$' can be replaced with '$r0 : r0$'.

In the above condition, simultaneous read and write of the same location is assumed to be supported. If this is not the case, then the $2PF2_a$ will consist only of FPs sensitized by simultaneous read operations; as a consequence the operations '$wx{:}rd$' in Condition $2PF2_a$ should be replaced with '$wy{:}n$', whereby n denotes no operation and $x \in \{0, 1\}$.

Conditions for detecting $2PF2_v$s

The $2PF2_v$ has the property that the fault is sensitized in the v-cell by first bringing the a-cell in a certain state, and thereafter applying two simultaneous operations to the v-cell. It consists of two FFMs: wCFrd&wRDF and wCFdrd&wDRDF; each with four FPs (see Table 6.4). In order to detect the presence of such faults in the v-cell c_v, the following condition has to be satisfied:

Condition 2PF2$_v$: Any wCFrd&wRDF and wCFdrd&wDRDF is detectable by a march test if the test selects all pairs of cells (c_a, c_v) whereby $a \in \{0, 1, ..., v - 1, v + 1, ..., n - 2, n - 1\}$; and each pair undergoes the four states 00, 01, 10 and 11. In addition, in each state two simultaneous read operations, followed by at least a single read operation, have to be applied to the v-cell.

Condition $2PF2_v$ can be explained as follows. When the pair of cells (c_a, c_v) is in state 00, then the wCFrd&wRDF FP '$< 0; 0r0 : 0r0/1/1 >_{a,v}$' will be sensitized and detected by simultaneous read operations applied to the v-cell; while the wCFdrd&wDRDF FP '$< 0; 0r0 : 0r0/1/0 >_{a,v}$' will be sensitized. The extra (single) read operation will detect the last fault. A similar explanation can be given for the other three states, which are required to detect the other six FPs of wCFrd&wRDF and wCFdrd&wDRDF.

Conditions for detecting $2PF2_{av}$s

The $2PF2_{av}$ has the property that the fault is sensitized in the v-cell by applying two simultaneous operations: one to the a-cell and one to the v-cell. It consists of three

FFMs: wCFds&wRDF, wCFds&wIRF, and wCFds&wRRF; each with four FPs (see Table 6.4). Note also here that in the table, e.g., the '$< dw1 : 0r0/1/1 >_{a,v}$' denotes only one FPs since d is the don't care value. The write operation simultaneously with the read operation will sensitize the fault irrespectively of the initial content of the cell being written. In order to detect the presence of such faults in cell c_v, we have to:

1. select all pairs (c_a, c_v) whereby $a \in \{0, 1, ..., v-1, v+1, ..., n-2, n-1\}$,

2. apply sensitizing operations to the two cells,

3. read the v-cell c_v.

The order in which c_v will be selected is not important, the only requirement is that v has take on all values from the set $\{0, 1, 2, ..., n-2, n-1\}$. The select order can be given as follows: $\Updownarrow_{v=0}^{n-1}$. In addition, the order in which c_a will be selected is also not important; the only requirement is that a has to take on all values from the set $\{0, 1, ..., v-1, v+1, ..., n-2, n-1\}$. Therefore, the select order can be given as follows: $\Updownarrow_{a=0}^{v-1}$ and $\Updownarrow_{a=v+1}^{n-1}$.

In the above, it is assumed that the a-cell, as well as the v-cell, can be any cell of the memory cell array. However, this is not the case in real designs. It has been shown in Section 5.3.2 using the inductive fault analysis and SPICE simulation, that the $2PF2_{av}$ can only be caused by bridges between bit lines belonging to different ports and to physical adjacent cells in the same row or column; see Figure 6.4. That means that if the a-cell is $c_a = c_{r,c}$ (i.e., a cell in row r and column c), then the v-cell has to be $c_{r\pm1,c}$ or $c_{r,c\pm1}$, such that the distance between a-cell and v-cell is just 1.

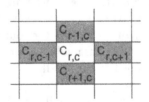

Figure 6.4. The v-cell space for a-cell$= c_{r,c}$

In addition, since the bridges between bit lines are to be found in metal layers (due to undesired extra metals), a defect causing a $2PF2_{av}$ between two cells (say c_1 and c_2) can be considered to be symmetrical with respect to the two cells. For example, if the application of the operation '$w0_{c1} : r1_{c2}$' (i.e., $w1$ applied to c_1 and $r1$ applied to c_2) causes a certain fault in c_2, then the same fault will be caused by applying the operation '$r1_{c1} : w0_{c2}$'. The only difference is that in the first case c_2 is the v-cell while in the second case c_1 is the v-cell. Therefore, for each a-cell $c_{r,c}$, one can restrict the v-cell space to one of the following:

a. $c_{r-1,c}$ and $c_{r,c-1}$,

b. $c_{r-1,c}$ and $c_{r,c+1}$,

c. $c_{r+1,c}$ and $c_{r,c-1}$, or

a. $c_{r+1,c}$ and $c_{r,c+1}$.

Note that, e.g., if for the a-cell=$c_{r,c}$ the two v-cells $c_{r+1,c}$ and $c_{r,c+1}$ are considered, then one does not need to consider $c_{r,c}$ as a v-cell when a-cell=$c_{r+1,c}$ (or $c_{r,c+1}$). This is because a defect between $c_{r,c}$ and $c_{r+1,c}$ is symmetrical with respects to the two cells. When a-cell=$c_{r+1,c}$, then the two cells $c_{r+2,c}$ and $c_{r+1,c+1}$ have to be considered as v-cells; see Figure 6.4. This reduction in a-cell and v-cell spaces has a significant impact on the condition to detect the $2PF2_{av}$ faults, and therefore on the test. It reduces the time complexity from $\Theta(n^2)$ to $\Theta(n)$. However, it requires the test to use topological addressing, rather than logical addressing [88].

Condition $2PF2_{av}$: Any wCFds&wRDF and wCFds&wIRF is detectable by a march test if the test contains one of the four pairs of march elements of Case A (e.g., the pair of march elements of Case A.1, Case A.2, Case A.3, or Case A.4), of Case B, of Case C, and of Case D. These four pairs of march elements can be combined into one, two, three, four, five, six, or seven march elements. In addition, a march test satisfying this condition can probabilistically detect wCFds&wRRF. That means that the detection of this fault can not be guaranteed due to the fact that the read operation produces a random value. Note that a nested addressing sequence is used; e.g., $\Uparrow_{c=0}^{C-1} \Uparrow_{r=0}^{R-1}$ whereby R and C denote the number of rows and columns in the memory cell array, respectively. The column c goes from 0 to $C-1$ ($\Uparrow_{c=0}^{C-1}$) or from $C-1$ to 0 ($\Downarrow_0^{c=C-1}$); and for each value of c, row r goes from 0 to $R-1$ ($\Uparrow_{r=0}^{R-1}$) or from $R-1$ to 0 ($\Downarrow_0^{r=R-1}$).

- Case A (to detect $< dw1 : 1r1/0/0 >_{a,v}$ and $< dw1 : 1r1/1/0 >_{a,v}$)

  A.1. $\Uparrow_{c=0}^{C-1} (\Uparrow_{r=0}^{R-1} (..., w1_{r,c} : r1_{r+1,c}, ...)); \Uparrow_{c=0}^{C-1} (\Uparrow_{r=0}^{R-1} (..., w1_{r,c} : r1_{r,c+1}, ...))$
  A.2. $\Uparrow_{c=0}^{C-1} (\Downarrow_0^{r=R-1} (..., w1_{r,c} : r1_{r-1,c}, ...)); \Uparrow_{c=0}^{C-1} (\Downarrow_0^{r=R-1} (..., w1_{r,c} : r1_{r,c+1}, ...))$
  A.3. $\Downarrow_0^{c=C-1} (\Uparrow_{r=0}^{R-1} (..., w1_{r,c} : r1_{r+1,c}, ...)); \Downarrow_0^{c=C-1} (\Uparrow_{r=0}^{R-1} (..., w1_{r,c} : r1_{r,c-1}, ...))$
  A.4. $\Downarrow_0^{c=C-1} (\Downarrow_0^{r=R-1} (..., w1_{r,c} : r1_{r-1,c}, ...)); \Downarrow_0^{c=C-1} (\Downarrow_0^{r=R-1} (..., w1_{r,c} : r1_{r,c-1}, ...))$

- Case B: (to detect $< dw1 : 0r0/1/1 >_{a,v}$ and $< dw1 : 0r0/0/1 >_{a,v}$)

  B.1. $\Uparrow_{c=0}^{C-1} (\Uparrow_{r=0}^{R-1} (..., w1_{r,c} : r0_{r+1,c}, ...)); \Uparrow_{c=0}^{C-1} (\Uparrow_{r=0}^{R-1} (..., w1_{r,c} : r0_{r,c+1}, ...))$
  B.2. $\Uparrow_{c=0}^{C-1} (\Downarrow_0^{r=R-1} (..., w1_{r,c} : r0_{r-1,c}, ...)); \Uparrow_{c=0}^{C-1} (\Downarrow_0^{r=R-1} (..., w1_{r,c} : r0_{r,c+1}, ...))$
  B.3. $\Downarrow_0^{c=C-1} (\Uparrow_{r=0}^{R-1} (..., w1_{r,c} : r0_{r+1,c}, ...)); \Downarrow_0^{c=C-1} (\Uparrow_{r=0}^{R-1} (..., w1_{r,c} : r0_{r,c-1}, ...))$
  B.4. $\Downarrow_0^{c=C-1} (\Downarrow_0^{r=R-1} (..., w1_{r,c} : r0_{r-1,c}, ...)); \Downarrow_0^{c=C-1} (\Downarrow_0^{r=R-1} (..., w1_{r,c} : r0_{r,c-1}, ...))$

- Case C: (to detect $< dw0 : 1r1/0/0 >_{a,v}$ and $< dw0 : 1r1/1/0 >_{a,v}$)

  C.1. $\Uparrow_{c=0}^{C-1} (\Uparrow_{r=0}^{R-1} (..., w0_{r,c} : r1_{r+1,c}, ...)); \Uparrow_{c=0}^{C-1} (\Uparrow_{r=0}^{R-1} (..., w0_{r,c} : r1_{r,c+1}, ...))$
  C.2. $\Uparrow_{c=0}^{C-1} (\Downarrow_0^{r=R-1} (..., w0_{r,c} : r1_{r-1,c}, ...)); \Uparrow_{c=0}^{C-1} (\Downarrow_0^{r=R-1} (..., w0_{r,c} : r1_{r,c+1}, ...))$

C.3. $\Downarrow_0^{c=C-1} (\Uparrow_{r=0}^{R-1} (..., w0_{r,c} : r1_{r+1,c}, ...)); \Downarrow_0^{c=C-1} (\Uparrow_{r=0}^{R-1} (..., w0_{r,c} : r1_{r,c-1}, ...))$

C.4. $\Downarrow_0^{c=C-1} (\Downarrow_0^{r=R-1} (..., w0_{r,c} : r1_{r-1,c}, ...)); \Downarrow_0^{c=C-1} (\Downarrow_0^{r=R-1} (..., w0_{r,c} : r1_{r,c-1}, ...))$

- Case D: (to detect $< dw0 : 0r0/1/1 >_{a,v}$ and $< dw0 : 0r0/0/1 >_{a,v}$)

  D.1. $\Uparrow_{c=0}^{C-1} (\Uparrow_{r=0}^{R-1} (..., w0_{r,c} : r0_{r+1,c}, ...)); \Uparrow_{c=0}^{C-1} (\Uparrow_{r=0}^{R-1} (..., w0_{r,c} : r0_{r,c+1}, ...))$

  D.2. $\Uparrow_{c=0}^{C-1} (\Downarrow_0^{r=R-1} (..., w0_{r,c} : r0_{r-1,c}, ...)); \Uparrow_{c=0}^{C-1} (\Downarrow_0^{r=R-1} (..., w0_{r,c} : r0_{r,c+1}, ...))$

  D.3. $\Downarrow_0^{c=C-1} (\Uparrow_{r=0}^{R-1} (..., w0_{r,c} : r0_{r+1,c}, ...)); \Downarrow_0^{c=C-1} (\Uparrow_{r=0}^{R-1} (..., w0_{r,c} : r0_{r,c-1}, ...))$

  D.4. $\Downarrow_0^{c=C-1} (\Downarrow_0^{r=R-1} (..., w0_{r,c} : r0_{r-1,c}, ...)); \Downarrow_0^{c=C-1} (\Downarrow_0^{r=R-1} (..., w0_{r,c} : r0_{r,c-1}, ...))$

Condition $2PF2_{av}$ can be explained as follows: the operation '$w1_{r,c} : r1_{r+1,c}$' in the first pair of march elements of Case A.1 will sensitize and detect the faults $< dw1 : 1r1/0/0 >_{a,v}$, $< dw1 : 1r1/1/0 >_{a,v}$, and may detect $< dw1 : 1r1/1/? >_{a,v}$, whereby the v-cell (i.e., $c_{r+1,c}$) and the a-cell ($c_{r,c}$) are adjacent in the same column; while the operation '$w1_{r,c} : r1_{r,c+1}$' will detect the same faults when the v-cell and the a-cell ($c_{r,c+1}$) are adjacent in the same row. It should be noted that in case $r = R$ (or $c = C$), then $r + 1$ should be replaced with $(r + 1)mod^2 R$ (and $c + 1$ with $(c + 1)mod\ C$). The other three pair of march elements of Case A detect the same faults as the first pair; however, they use different addressing orders. For example Case A.2 considers the a-cell $c_{r,c}$ and the two v-cells $c_{r-1,c}$ and $c_{r,c+1}$, etc. A similar explanation applies to Case B, Case C, and Case D.

6.4 Tests for two-port faults

The FFMs for 2P memory cell array faults are divided into 1PFs and 2PFs. The 1PFs are the conventional faults that can occur in SP memories; while the 2PFs are faults based on simultaneous operations applied to the 2P memory. Therefore, the test procedure can be divided into two parts:

1. Test(s) to detect 1PFs.

2. Test(s) to detect 2PFs.

Tests for 1PFs are described in literature like in [2, 21, 24, 88]; while tests for 2PFs have to be derived. Below, tests for detecting 2PFs will be introduced. First, tests for 2PF1s will be presented and thereafter for 2PF2s.

6.4.1 Tests for the 2PF1s

The test shown in Figure 6.5, referred as *March 2PF1*, detects all 2PF1 faults. March 2PF1 consists of three march elements: M_0, M_1, and M_2; it has a test length of $7n$, whereby n is the size of the memory cell array. Note that the single read operations in M_1 and M_2 can be replaced with simultaneous read operations; e.g., '$r1 : -$' can be replaced with '$r1 : r1$'.

---

[2] a mod b= the remainder of dividing a by b

$$\{ \; \updownarrow (w0:n) \; ; \; \updownarrow (w1:r0, r1:r1, r1:-) \; ;$$
$$M_0 \qquad\qquad\qquad M_1$$
$$\updownarrow (w0:r1, r0:r0, r0:-) \; \}$$
$$M_2$$

Figure 6.5. March 2PF1

March 2PF1 detects all 2PF1 faults since it satisfies Condition 2PF1 of Section 6.3.1. The second march element of the test (i.e., M_1) contains the three march elements of Case A.2 merged into one march element, while M_2 contains the three march elements of Case A.1. The three march elements of March 2PF1 can be merged into one or two elements without impacting the fault coverage.

It should be noted that simultaneous read and write operations of the same location (with read data discarded) is assumed to be allowed. If this is not supported, then the FFM wRDF&wTF is not realistic; and as a consequence, March 2PF1 can be simplified. The result is shown in Figure 6.6, and referred as *March 2PF1-*; it has a test length of $6n$. Note that the four march elements of the test can be merged into one, two, or three march elements.

$$\{ \; \updownarrow (w0:n) \; ; \; \updownarrow (r0:r0, r0:-) \; ;$$
$$M_0 \qquad\qquad M_1$$
$$\updownarrow (w1:-) \; ; \; \updownarrow (r1:r1, r1:-) \; \}$$
$$M_2 \qquad\qquad M_3$$

Figure 6.6. March 2PF1-

6.4.2 Tests for the 2PF2s

The 2PF2s are divided, depending on the cells to which the simultaneous operations are applied, into three types: $2PF2_a$, $2PF2_v$, and $2PF2_{av}$. In the following, tests for each type will be introduced.

Tests for the $2PF2_a$ faults

March $2PF2_a$, shown in Figure 6.7, detects all $2PF2_a$ faults since it satisfies Condition $2PF2_a$ of Section 6.3.2: M_1 and M_2 contain the two march elements of Case A, while M_3 and M_4 contain the two march elements of Case B. Note that the operations '$r0:-$' and '$r1:-$' in the test can be replaced with '$r0:r0$', respectively, with '$r1:r1$'. March $2PF2_a$ has a test length of $12n$, and can detect the 2PF1s: wRDF&wRDF and wRDF&wTF; see Table 6.4. The wRDF&wRDF will be detected since March $2PF2_a$ satisfies Condition wRDF&wRDF of Section 6.3.1: M_1 contains the march elements of Case A, while M_3 contains march elements of Case B. On the other hand, March $2PF2_a$ also satisfies Condition wRDF&wTF of Section 6.3.1: M_1 (also M_3 and M_4) contains the pair of march elements of Case A merged into one single march element, while M_1 and M_2 (also M_3) contain the pair of march elements of Case B.

$$\{ \ \updownarrow (w0:n) \ ; \\ \qquad M_0 \\ \updownarrow (r0:r0, w1:r0, r1:r1, w0:r1) \ ; \ \updownarrow (r0:-, w1:-) \ ; \\ \qquad M_1 \qquad\qquad\qquad\qquad\qquad M_2 \\ \updownarrow (r1:r1, w0:r1, r0:r0, w1:r0) \ ; \ \updownarrow (r1:-) \ \} \\ \qquad M_3 \qquad\qquad\qquad\qquad M_4$$

Figure 6.7. March $2PF2_a$

March $2PF2_a$ can be further optimized to a $10n$ test without impacting the fault coverage. The result is shown in Figure 6.8, and referred as *March 2PF2$_a$-*. Note that the addressing sequence of the optimized version is *relevant*; it is important for the fault coverage as it will be shown below. Although the test of Figure 6.8 does not explicitly contain the two pairs of march elements of Condition $2PF2_a$, it detects all $2PF2_a$ faults; the sensitizing operations for $2PF2_a$ faults are distributed over M_1, M_2, M_3, M_4, and M_5.

$$\{ \ \updownarrow (w0:n) \ ; \ \Uparrow (r0:r0, w1:r0) \ ; \ \Uparrow (r1:r1, w0:r1) \\ \qquad M_0 \qquad\qquad M_1 \qquad\qquad\qquad M_2 \\ \Downarrow (r0:r0, w1:r0) \ ; \ \Downarrow (r1:r1, w0:r1) \ ; \ \Downarrow (r0:-) \ \} \\ \qquad M_3 \qquad\qquad\qquad M_4 \qquad\qquad M_5$$

Figure 6.8. March $2PF2_a$-

The $2PF2_a$ faults consists of one FFM, with eight FPs; see Table 6.4.

- The $< 0r0 : 0r0; 0/1/- >_{a,v}$ and the $< dw1 : drd; 0/1/- >_{a,v}$ will be sensitized and detected by M_1 of Figure 6.8 if the v-cell has a higher address than the a-cell; i.e., $v > a$. If $v < a$, then these faults will be sensitized and detected by M_3.

- The $< 1r1 : 1r1; 1/0/- >_{a,v}$ and $< dw0 : drd; 1/0/- >_{a,v}$ will be sensitized and detected by M_2 if $v > a$. If $v < a$, then these faults will be sensitized and detected by M_4.

- The $< 0r0 : 0r0; 1/0/- >_{a,v}$ and the $< dw1 : drd; 1/0/- >_{a,v}$ will be sensitized by M_1 and detected by M_2 if $v < a$; while the same faults will be sensitized by M_3 and detected by M_4 if $v > a$.

- The $< 1r1 : 1r1; 0/1/- >_{a,v}$ and the $< dw0 : drd; 0/1/- >_{a,v}$ will be sensitized by M_2 and detected by M_3 if $v < a$; while the same faults will be sensitized by M_4 and detected by M_5 if $v > a$.

Again, if simultaneous read and write is not supported, then the operations, e.g., '$w0 : r1$' in March 2PF2$_a$ (and March 2PF2$_a$-) should be replaced with '$w1 : n$'. In that case, the 2PF2$_a$ will consist only of FPs based on two simultaneous read operations.

Tests for the 2PF2$_v$ faults

The test detecting all 2PF2$_v$ faults is shown in Figure 6.9, and is referred as *March 2PF2$_v$*; it satisfies Condition 2PF2$_v$ of Section 6.3.2. M_0 initializes all memory cells to 0; which means that the state of all pairs (c_a, c_v) is 00. In the second march element, the v-cell is first read via the two ports simultaneously while the state of all pairs (c_a, c_v) is 00; therefore, the FP '$< 0; 0r0 : 0r0/1/1 >_{a,v}$' (of wCFrd&wRDF) will be detected, while the FP '$< 0; 0r0 : 0r0/1/0 >_{a,v}$' (of wCFdrd&wDRDF) will be sensitized (see Table 3). The latter will be detected by the next operation within the same march element. Thereafter, the v-cell will be written with 1, and then will be read via the two ports simultaneously. This means that simultaneous read operations are applied to the v-cell while the state of all pairs (c_a, c_v) is 01; therefore the FP '$< 0; 1r1 : 1r1/0/0 >_{a,v}$' will be detected, while the FP '$< 0; 1r1 : 1r1/0/1 >_{a,v}$' will be sensitized. The latter fault will be detected by the next read operation within the same march element. Finally, the v-cell is written with 0, such that all pairs (c_a, c_v) again enter state 00. A similar explanation can be given for march elements M_2 and M_3. Note that the operations '$r0 : -$' and '$r1 : -$' in M_1 and M_3 can be replaced with '$r0 : r0$', respectively, '$r1 : r1$'. March 2PF2$_v$ has a test length of $14n$, and also detects the 2PF1 faults: wDRDF&wDRDF and wRDF&wRDF (see also Figure 6.5 and Table 6.4).

$$
\begin{array}{c}
\{\ \Updownarrow (w0 : n)\ ;\ \Updownarrow (r0 : r0, r0 : -, w1 : -, r1 : r1, r1 : -, w0 : -)\ ; \\
M_0 \qquad\qquad\qquad\qquad M_1 \\
\Updownarrow (w1 : -)\ ;\ \Updownarrow (r1 : r1, r1 : -, w0 : -, r0 : r0, r0 : -, w1 : -)\ \} \\
M_2 \qquad\qquad\qquad\qquad M_3
\end{array}
$$

Figure 6.9. March 2PF2$_v$

March 2PF2$_v$ can also be optimized without impacting the fault coverage of the 2PF2$_v$ faults. The result is shown in Figure 6.10, and referred as *March 2PF2$_v$-*. It consists of five march elements, and has a test length of $13n$. Note that the addressing sequence of the optimized version is *relevant*; it is important for the fault coverage as it will be shown below. Note also that the operations '$r0 : -$' and '$r1 : -$' in the test can be replaced with '$r0 : r0$', respectively, '$r1 : r1$'.

$$\{ \updownarrow (w0:n) \ ;$$
$$M_0$$
$$\Uparrow (r0:r0,r0:-,w1:-) \ ; \quad \Uparrow (r1:r1,r1:-,w0:-) \ ;$$
$$M_1 \qquad\qquad\qquad\qquad M_2$$
$$\Downarrow (r0:r0,r0:-,w1:-) \ ; \quad \Downarrow (r1:r1,r1:-,w0:-) \ \}$$
$$M_3 \qquad\qquad\qquad\qquad M_4$$

Figure 6.10. March $2PF2_v-$

March $2PF2_v$- detects all $2PF2_v$ faults since Condition $2PF2_v$ of Section 6.3.2 is satisfied. Table 6.5 shows the operations performed on two cells c_i and c_j by the march elements of Figure 6.10. The table contains a column 'state', which identifies the state $S_{i,j}$ of the two cells (c_i, c_j) *before* the operation is performed; and a column 'State $S_{i,j}$', which identifies the state *after* the operation. The table shows that all states of (c_i, c_j) (i.e., 00, 01, 11, 10) are generated, and in each state two simultaneous read operations followed by (at least) a single read operation are applied to cell c_i and c_j.

Tests for the $2PF2_{av}$ faults

The test shown in Figure 6.11, referred to as *March $2PF_{av}$*, detects all wCF$_{ds}$&wRDF faults, all wCF$_{ds}$&wIRF faults, and also probabilistically detects wCF$_{ds}$&wRRF faults since it satisfies Condition $2PF2_{av}$ of Section 6.3.2: M_1 contains the first march element of Case B.1 and of Case D.1; M_2 contains the second march element of Case B.1 and of Case D.1. In addition, M_4 contains the first march element of Case C.1 and of Case A.1; while M_5 contains the second march elements of Case C.1 and of Case A.1. March $2PF_{av}$ has a test length of $10n$.

$$\{ \updownarrow (w0:n) \ ; \quad \Uparrow_{c=0}^{C-1} (\Uparrow_{r=0}^{R-1} (w1_{r,c}:r0_{r+1,c}, w0_{r,c}:r0_{r+1,c})) \ ;$$
$$M_0 \qquad\qquad\qquad\qquad M_1$$
$$\Uparrow_{c=0}^{C-1} (\Uparrow_{r=0}^{R-1} (w1_{r,c}:r0_{r,c+1}, w0_{r,c}:r0_{r,c+1})) \ ;$$
$$M_2$$
$$\updownarrow (w1:-) \ ; \quad \Uparrow_{c=0}^{C-1} (\Uparrow_{r=0}^{R-1} (w0_{r,c}:r1_{r+1,c}, w1_{r,c}:r1_{r+1,c})) \ ;$$
$$M_3 \qquad\qquad\qquad\qquad M_4$$
$$\Uparrow_{c=0}^{C-1} (\Uparrow_{r=0}^{R-1} (w0_{r,c}:r1_{r,c+1}, w1_{r,c}:r1_{r,c+1})) \ \}$$
$$M_5$$

Figure 6.11. March $2PF2_{av}$ (= March d2PF)

Below, an optimal version of March $2PF2_{av}$ will be given. The test, referred to as *March $2PF2_{av}$-*, is shown in Figure 6.12; it has a test length of $9n$ and consists of nine march elements, each with only one pair of operations. It can be clearly seen that the test, similar to March $2PF2_{av}$, covers the $2PF2_{av}$ faults since it satisfies Condition

Table 6.5. State table for detecting $2PF2_v$ faults

Step	March element	State	Operation	State $S_{i,j}$
1	M_0	--	'$w0 : n$' to c_i	0-
2		0-	'$w0 : n$' to c_j	00
3	M_1	00	'$r0 : r0$' to c_i	00
4		00	'$r0 : -$' to c_i	00
5		00	'$w1 : -$' to c_i	10
6		10	'$r0 : r0$' to c_j	10
7		10	'$r0 : -$' to c_j	10
8		10	'$w1 : -$' to c_j	11
9	M_2	11	'$r1 : r1$' to c_i	11
10		11	'$r1 : -$' to c_i	11
11		11	'$w0 : -$' to c_i	01
12		01	'$r1 : r1$' to c_j	01
13		01	'$r1 : -$' to c_j	01
14		01	'$w0 : -$' to c_j	00
15	M_3	00	'$r0 : r0$' to c_j	00
16		00	'$r0 : -$' to c_j	00
17		00	'$w1 : -$' to c_j	01
18		01	'$r0 : r0$' to c_i	01
19		01	'$r0 : -$' to c_i	01
20		01	'$w1 : -$' to c_i	11
21	M_4	11	'$r1 : r1$' to c_j	11
22		11	'$r1 : -$' to c_j	11
23		11	'$w0 : -$' to c_j	10
24		10	'$r1 : r1$' to c_i	10
25		10	'$r1 : -$' to c_i	10
26		10	'$w0 : -$' to c_i	00

$2PF2_{av}$ of Section 6.3.2: M_1, M_2, M_3, and M_4 contain the first march elements of Case B.1, Case A.1, Case C.1, and Case D.1, respectively; while M_5, M_6, M_7, and M_8 contain the second march elements of Case B.1, Case A.1, Case C.1, and Case D.1, respectively.

Another optimal version of March $2PF2_{av}$ is shown in Figure 6.13. It also has a test length of $9n$, but it consists only of three march elements. The test satisfies Condition $2PF2_{av}$ of Section 6.3.2 as follows: M_2 contains the first march elements of Case B.1, Case A.1, Case C.1, and Case D.1; while M_3 contains the second march elements of Case B.1, Case A.1, Case C.1, and Case D.1. Note that the read and the write operations are interchanged in some cases; e.g., '$w1 : r0$' changed into '$r0 : w1$'. This has no negative impact on the fault coverage, since a defect causing a $2PF2_{av}$ between two cells is symmetrical with respect to the two cells (see Condition for detecting $2PF_{av}$s in Section 6.3.2. That means that if a fault sensitized by '$w1_{c1}$:

$$\{ \; \Updownarrow (w0:n) \; ; \; \Uparrow_{c=0}^{C-1} (\Uparrow_{r=0}^{R-1} (w1_{r,c} : r0_{r+1,c})) \; ; \; \Uparrow_{c=0}^{C-1} (\Uparrow_{r=0}^{R-1} (w1_{r,c} : r1_{r+1,c})) \; ;$$
$$M_0 \hspace{3.5cm} M_1 \hspace{3.5cm} M_2$$
$$\Uparrow_{c=0}^{C-1} (\Uparrow_{r=0}^{R-1} (w0_{r,c} : r1_{r+1,c})) \; ; \; \Uparrow_{c=0}^{C-1} (\Uparrow_{r=0}^{R-1} (w0_{r,c} : r0_{r+1,c})) \; ;$$
$$M_3 \hspace{3.5cm} M_4$$
$$\Uparrow_{c=0}^{C-1} (\Uparrow_{r=0}^{R-1} (w1_{r,c} : r0_{r,c+1})) \; ; \; \Uparrow_{c=0}^{C-1} (\Uparrow_{r=0}^{R-1} (w1_{r,c} : r1_{r,c+1})) \; ;$$
$$M_5 \hspace{3.5cm} M_6$$
$$\Uparrow_{c=0}^{C-1} (\Uparrow_{r=0}^{R-1} (w0_{r,c} : r1_{r,c+1})) \; ; \; \Uparrow_{c=0}^{C-1} (\Uparrow_{r=0}^{R-1} (w0_{r,c} : r0_{r,c+1})) \; \}$$
$$M_7 \hspace{3.5cm} M_8$$

Figure 6.12. March $2PF2_{av}$- (= March d2PF-); optimal version 1

$r0_{c2}$' (i.e., $w1$ applied to c_1 and $r1$ applied to c_2), then the same fault will also be sensitized with '$r0_{c1} : w1_{c2}$'; the only difference is that in the first case c_2 is the v-cell (i.e, the cell where the fault effect appears), while in the second case c_1 is the v-cell.

$$\{ \; \Updownarrow (w0:n) \; ; \; \Uparrow_{c=0}^{C-1} (\Uparrow_{r=0}^{R-1} (w1_{r,c} : r0_{r+1,c}, r1_{r,c} : w1_{r+1,c}, w0_{r,c} : r1_{r+1,c}, r0_{r,c} : w0_{r+1,c})) \; ;$$
$$M_0 \hspace{6cm} M_1$$
$$\Uparrow_{c=0}^{C-1} (\Uparrow_{r=0}^{R-1} (w1_{r,c} : r0_{r,c+1}, r1_{r,c} : w1_{r,c+1}, w0_{r,c} : r1_{r,c+1}, r0_{r,c} : w0_{r,c+1})) \; \}$$
$$M_2$$

Figure 6.13. March $2PF2_{av}$- (= March d2PF-); optimal version 2

6.4.3   Classification of 2PF tests

The proposed march tests for 2P memories can be classified, based on the type of addressing they use, into two classes:

- *Single-addressing tests*: these are tests which access one cell at a time (i.e., both ports use the same address). They consist of March 2PF1, March $2PF2_a$ and March $2PF2_v$.

- *Double-addressing tests*: these are tests which access two different locations at a time. They consist only of March $2PF2_{av}$. From here on, this test will be referred to as *March d2PF* (d stands for double addressing).

The three single-addressing tests can be merged into a single march test; the result is shown in Figure 6.14 and is referred as *March s2PF* (s for single addressing). The test satisfies Condition 2PF1 (of Section 6.3.1) required to detect 2PF1s; it also satisfies Condition $2PF2_a$ and Condition $2PF2_v$ of Section 6.3.2.

- Condition 2PF1: satisfied by M_1 and M_2, as well as by M_4 and M_5.

- Condition $2PF2_a$: satisfied by M_1, M_2, M_4, and M_5. M_1 and M_2 contain the two march elements of Case A; while M_4 and M_5 contain the two march elements of Case B.

- Condition $2PF2_v$: satisfied by M_1 and M_4. Note that M_1 and M_4 are extended versions of M_1 and M_3 of March $2PF2_v$; see Figure 6.9.

It has to be clear from Figure 6.14 that March s2PF has a test length of $16n$; while the test lengths of March 2PF1, March $2PF2_a$ and March $2PF2_v$ are $7n$, $12n$, and $14n$, respectively. Therefore, in order to detect 2PF1, $2PF2_a$ and $2PF2_v$ faults, one can use March s2PF instead of testing for these faults separately. This will reduce the test time with 51.15%; i.e., from $7n + 12n + 14n = 33n$ to $16n$.

$$
\begin{aligned}
&\{ \; \Updownarrow (w0:-) \; ; \\
&\qquad M_0 \\
&\Updownarrow (r0:r0, r0:-, w1:r0, r1:r1, r1:-, w0:r1) \; ; \; \Updownarrow (r0:-) \; ; \\
&\qquad\qquad\qquad\qquad\qquad M_1 \qquad\qquad\qquad\qquad\qquad\qquad M_2 \\
&\Updownarrow (w1:-) \; ; \\
&\qquad M_3 \\
&\Updownarrow (r1:r1, r1:-, w0:r1, r0:r0, r0:-, w1:r0) \; ; \; \Updownarrow (r1:-) \; \} \\
&\qquad\qquad\qquad\qquad\qquad M_4 \qquad\qquad\qquad\qquad\qquad\qquad M_5
\end{aligned}
$$

Figure 6.14. March s2PF for 2PF1s, $2PF2_a$s and $2PF2_v$s

March s2PF can be further optimized, without impacting the fault coverage, by splitting M_1 and M_4 each into two march elements. Figure 6.15 shows the result, referred to as *March s2PF-*. It consists of six march elements and has a test length of $14n$; i.e., $2n$ less than March s2PF. That means that if this version is used to detect 2PF1, $2PF2_a$ and $2PF2_v$ faults, then the test time reduction will be 57.57% (i.e., from $33n$ to $14n$). Note that the addressing sequence of the optimized version is *relevant*; it is important for the fault coverage.

$$
\begin{aligned}
&\{ \; \Updownarrow (w0:n) \; ; \\
&\qquad M_0 \\
&\Uparrow (r0:r0, r0:-, w1:r0) \; ; \; \Uparrow (r1:r1, r1:-, w0:r1) \; ; \\
&\qquad\qquad M_1 \qquad\qquad\qquad\qquad\qquad M_2 \\
&\Downarrow (r0:r0, r0:-, w1:r0) \; ; \; \Downarrow (r1:r1, r1:-, w0:r1) \; ; \\
&\qquad\qquad M_3 \qquad\qquad\qquad\qquad\qquad M_4 \\
&\Downarrow (r0:-) \; \} \\
&\qquad M_5
\end{aligned}
$$

Figure 6.15. March s2PF-

March s2PF- detects all 2PF1s since it satisfies Condition 2PF1 of Section 6.3.1: M_1 and M_2 (also M_3 and M_4) contain the pair B.1, while M_2 and M_3 (also M_4 and M_5) contain the pair B.2. In addition, it detects all $2PF_a$ faults. Note that the march elements of March s2PF2- are the same as those of March $2PF2_a$- (see Figure 6.8);

except M_1, M_2, M_3, and M_4 are extended with read operations (that do not impact the fault coverage). Moreover all $2PF2_v$ faults will be detected since M_1 through M_4 of March s2PF- are the same as those of March $2PF2_v$ (see Figure 6.10), with the only difference that '$w0:-$' in M_2 and M_4 is replaced with '$w0:r1$', and '$w1:-$' in M_1 and M_3 with '$w1:r0$'; that does not impact the fault coverage.

It should be noted that for March s2PF and for March s2PF-, simultaneous read and write of the same location has to be allowed. If this is not the case, then all operations '$wx:ry$' in Figure 6.14 and Figure 6.15 should be replaced with '$wx:n$'; whereby $x, y \in \{0, 1\}$.

6.4.4 Summary of 2P tests

Table 6.6 summarizes the tests introduced in this section. It shows their required number of operations (i.e., test length (T.L.)) including the initialization, together with their fault coverage; see also Table 6.4.

Table 6.6. Summary of the 2P tests

Test	T.L.	Fault coverage
March 2PF1	$7n$	All 2PF1s
March $2PF2_a$	$12n$	All $2PF2_a$s The 2PF1s: wRDF&wRDF and wRDF&wTF
March $2PF2_a$-	$10n$	All $2PF2_a$s The 2PF1s: wRDF&wRDF, wRDF&wTF
March $2PF2_v$	$14n$	All $2PF2_v$s The 2PF1s: wDRDF&wDRDF and wRDF&wRDF
March $2PF2_v$-	$13n$	All $2PF2_v$s The 2PF1s: wDRDF&wDRDF and wRDF&wRDF
March d2PF	$10n$	All $2PF2_{av}$s
March d2PF2-	$9n$	All $2PF2_{av}$s
March s2PF	$16n$	All 2PF1s, all $2PF2_a$s, all $2PF2_v$s
March s2PF-	$14n$	All 2PF1s, all $2PF2_a$s, all $2PF2_v$s

Based on the above, one can conclude that the test for 2PFs in 2P memories can be done using two *linear* march tests:

1. March s2PF (with a test length of $16n$), or March s2PF- ($14n$); and

2. March d2PF ($10n$), or March d2PF- ($9n$).

6.5 Comparison with other tests

This section gives an analytical comparison of March s2PF- and March d2PF- (which are sufficient for detecting all 2PFs) with other industrial 2P tests (which are mainly

based on the extension of the conventional tests designed for single-port memories).
Here, three single-port memory tests will be considered and extended to dual-port
tests: Scan [2], March C- [49, 88], PMOVI [39]; these tests are extended by replacing
the single read operation with two simultaneous read operations. For example:

Scan=$\{\Uparrow (w0); \Uparrow (r0); \Uparrow (w1); \Uparrow (r1)\}$ is changed into dual-port Scan as:

2P-Scan = $\{\Uparrow (w0 : n); \Uparrow (r0 : r0); \Uparrow (w1 : n); \Uparrow (r1 : r1)\}$.

In addition, the two test algorithms presented by [100] for dual-port memories
will be considered; they are called MMCA and WIPD:

MMCA=$\{\Uparrow_{i=0}^{n-1} (w0_i);$
$\Uparrow_{i=0}^{n-1} (r0_i : rx_{i-2}, w1_i : rx_{i+2});$
$\Uparrow_{i=0}^{n-1} (r1_i : rx_{i-2}, w0_i : rx_{i+2});$
$\Downarrow_{i=0}^{n-1} (r0_i : rx_{i-2}, w1_i : rx_{i+2});$
$\Downarrow_{i=0}^{n-1} (r1_i : rx_{i-2}, w0_i : rx_{i+2});$
$\Updownarrow_{i=0}^{n-1} (r0_i : n)\}$

where 'rx' denotes a read data 'x" without observing it, and 'n' denotes no operation.

WIPD=$\{\Uparrow_{i=0}^{n-1} (w0_i);$
$\Uparrow_{i=2-1}^{n-1} (w0_i : w1_{i-2}, r0_{i-1} : n); \Uparrow_{i=0}^{n-3} (r1_{i-1} : n);$
$\Uparrow_{i=2-1}^{n-1} (w1_i : w0_{i-2}, r1_{i-1} : n); \Uparrow_{i=0}^{n-3} (r0_{i-1} : n)\}$

Table 6.7 summarizes the fault coverage of all the 2P tests considered here; all
tests have a linear time complexity. The test length (T.L.) of each test is also given;
n denotes the size of the memory. In the table, "a/b" denotes that the test detects
'a' of the 'b' FPs of the correspondent fault subclass. E.g., 2P-March C- detects
two FPs of the total of six FPs that 2PF1 fault class consists of; see also Table 6.4.
The table clearly shows that using the first five tests (i.e., 2P-Scan, 2P-March C-,
2P-PMOVI, MMCA and WIPD), the fault coverage of the targeted unique 2PFs will
be not 100%. This is well the case by using March s2PF- and March d2PF-.

Table 6.7. Comparison of the dual-port tests

Tests	T. L.	2PF1	2PF2$_a$	2PF2$_v$	2PF2$_{av}$
2P-Scan	$4n$	2/6	1/8	2/8	0/8
2P-March C-	$10n$	2/6	4/8	4/8	0/8
2P-PMOVI	$13n$	4/6	4/8	8/8	0/8
MMCA	$10n$	0/6	0/8	0/8	4/8
WIPD	$7n$	0/6	0/8	0/8	0/8
March s2PF-	$14n$	6/6	8/8	8/8	0/8
March d2PF-	$9n$	0/6	0/8	0/8	8/8

6.6   Test strategy

The realistic FFMs for 2P memories have been derived and divided into 1PFs and
2PFs. Many tests for 1PFs are know in the literature like in [2, 21, 24, 88], and
other ones are introduced in Section 6.2; while tests for 2PFs have been developed
in Section 6.4. Now the test strategy can be established. Figure 6.16 show the test
strategy for 2P memory (with two read-write ports). First, the 1PFs are tested by
applying SP tests; for instance March SS [30]. Note that it is necessary to apply the
test through each port separately. This is because the 1PFs can be *cell faults* or *port
faults*; whereby a cell fault is a fault caused by a defect within a *memory cell* (e.g.,
an open at the pull down of the cell), while a port fault is a fault caused by a defect
related to a *certain port* (e.g., an open in the bit line of port P_a). The cell faults can
be thus tested just via one port; however, port faults require the test to be applied
via each port separately.

After testing 1PFs, the dies may and may not pass the applied SP tests. If the
die fails the SP test(s), then it does not make sense to test it for 2PFs. However, if
the die passes the SP test(s), the application of the 2P tests is required. It has been
shown in the Section 6.4.4, that all 2PFs can be tested by March s2PF- and March
d2PF-. Therefore, one can only use these two tests to detect the 2PFs for the dies
passing the SP test(s). Note that during the test, if a die fails any of the tests, then
the test procedure will be stopped with as a result 'Die fails'; no additional test is
required in that case. Assuming that only March SS is applied to detect 1PFs, the
test strategy of Figure 6.16 requires a test length of $67n$ since:
- March SS via P_a requires $22n$,
- March SS via P_b requires $22n$,
- March s2PF- via the two ports requires $14n$, and
- March d2PF- via the two ports requires $9n$.

It should be noted that when March d2PF- is designed, it is assumed that the
defects causing $2PF2_{av}$ faults between two cells (mainly bridges between bit lines
belonging to different ports) are symmetrical with respect to the two cells. However,
one can ignore this assumption and apply March d2PF- *two times* by reversing the
two ports.

6.7   Test results versus fault probabilities

In Section 5.4, the probabilities of 1PFs and 2PFs have been determined for two
2P memories having two different layouts; namely ML and WL[3]. The two layouts
implement the same electrical memory circuit. The probabilities have been calculated
first by by assuming that all SDs are equal likely to occur, and thereafter by using

---

[3]ML and WL are two Intel designs

```
Detection of 1PFs:
    Apply single port test(s) (e.g., March SS) via port Pₐ;
    If die fails, go to FAIL;
    Apply single port test(s) (e.g., March SS) via port P_b;
    If die fails, go to FAIL;
Detection of 2PFs:
    Apply March s2PF- via the two ports;
    If die fails, go to FAIL;
    Apply March d2PF- via the two ports;
    If die fails, go to FAIL;
Pass: Print 'Die passes';
       END;
Fail: Print 'Die fails';
       END;
```

Figure 6.16. Test strategy for 2P memories

Inductive Fault Analysis (IFA). The results found using IFA are given in Table 5.9, and are summarized below:

- For ML layout:
    - 1PFs consist of 94.409%
    - 2PFs consist 5.591%

- For WL layout:
    - 1PFs consist of 98.613%
    - 2PFs consist of 1.387%

Note that the percentage of 2PFs for WL is $\frac{5.591}{1.387} \simeq 4.03$ times smaller than for ML. Therefore, it is expected that more ML dies will fail the 2P tests than WL dies.

In order to industrially evaluate the 2P tests introduced in Section 6.4, a version of the test strategy of Figure 6.16 has been applied to the ML and WL designs. That work has been done at Intel Corporation, CA, USA. For the detection of 1PFs, a large set of SP tests have been used (e.g., Scan, MATS+, March C-, etc). The dies passing these are thereafter tested using versions of March s2PF- and March d2PF-. The preliminary test results[4] show the following:

- For ML design: From 33830 ML dies, passing all used SP tests, 23 dies failed to pass the implemented 2P tests; 7 failed to pass March s2PF-, and 21 failed

---

[4]Note: These results give just an idea about the effectiveness of the new 2PF tests; they do not present high volume production

to pass March d2PF- (note that 5 dies failed to pass both tests). That means that the tests detect 0.0678% of the dies passing all SP tests; which corresponds with a reduction of of 680 Defects per Million (DPM).

- For WL design: From 2165868 WL dies, passing all used SP test, 305 dies fail for 2P March s2PF- and March d2PF-; which corresponds with a reduction of 141 DPM. That is $\frac{680}{141} \simeq 4.82$ times smaller than DPM level for ML.

It is interesting to note that the predicted ratios of 2PFs, based on IFA, is 4.03, while the fault coverage ratio of the 2P tests for ML and WL is 4.82. This shows that the predicted ratio of 4.03 (based on IFA) reasonably resembles the real fault coverage ratio of 4.83 with an accuracy of about 83.43%.

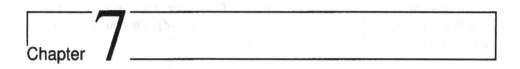

## Chapter 7

Testing restricted two-port SRAMs

7.1 Classification of two-port memories

7.2 Realistic faults for restricted two-port memories

7.3 Tests for restricted two-port memories

7.4 Test strategy for restricted two-port memories

In Chapter 5 and Chapter 6, realistic fault models based on defect injection and circuit simulation, respectively test algorithms, for two-port (2P) SRAMs have been introduced. However, these were only valid for 2P memories whereby both ports have read as well as write capabilities. The 2P memories can have port restrictions; i.e., the port may only allow for read or write operations. These restrictions impact the possible fault models, and hence also the tests. The test strategy, which determines which set of single-port (SP) tests and 2P tests to be performed for an optimal fault coverage, is also impacted by port restrictions.

In this chapter, first a classification of 2P memories will be done, based on the type of the port they consists of; read-only, write only, or read-write port. Then, the impact of port restrictions on the fault modeling will be stated, and realistic fault models for each 2P memory class will be presented. Thereafter, tests detecting such faults will be established. Finally, the test strategy for each class will be covered.

7.1 Classification of two-port memories

2P memories come in different forms, depending on the type of ports they consist of. Each of the two ports may have the capability to be a *read-only port (Pro)*, a *write-only port (Pwo)*, or a *read-write port (Prw)*. The total number of ports $p = \#Prw + \#Pwo + \#Pro$ (Note: # denotes the number of). Therefore, four types of 2P memories can be distinguished based on the *port mix*:

- (rw-rw)2P memories with $\#Prw = 2$, $\#Pro = 0$, and $\#Pwo = 0$.

- (rw-ro)2P memories with $\#Prw = 1$, $\#Pro = 1$, and $\#Pwo = 0$.

- (wo-rw)2P memories with $\#Prw = 1$, $\#Pro = 0$, and $\#Pwo = 1$.

- (wo-ro)2P memories with $\#Prw = 0$, $\#Pro = 1$ and $\#Pwo = 1$

The name of each type is chosen in the sense of indicating implicitly the port number; e.g., for (wo-rw)2P memories, the first port P_a=Pwo and the second port P_b=Prw.

Depending on the number of simultaneous read operations they allow for, the above four types of 2P memories can be further classified into two classes:

- *Two-read 2P memories*: these are memories with two ports both having the read capability. Therefore, the memories support two simultaneous read operations to the same address. This class consist of (rw-rw)2P and (rw-ro)2P memories.

- *Single-read 2P memories*: these are memories with two ports, but only one of the ports has the read capability. Therefore, only one read operation is allowed at a time. This class consists of (rw-wo)2P and (ro-wo)2P memories.

The functional fault model (FFMs) derived in Chapter 5 are only valid for (rw-rw)2P memories. In the next section, the impact of port restrictions on these FFMs will be established.

7.2 Realistic faults for restricted 2P memories

FFMs for all 2P memory types can, similar to those of (rw-rw)2P memories, be divided into 1PFs and 2PFs. The 2PFs are *port mix dependent* since they require two simultaneous operations in order to be sensitized; e.g., a FFM based on two simultaneous read operations can not occur in a (wo-ro)2P memory. The consequences of port restrictions on the 2PFs of Table 5.6 are explained in the subsections below. This will be done for each 2P memory class separately.

7.2.1 Realistic 2PFs for two-read 2P memories

Depending on their design, two-read 2P memories may and may not support simultaneous read and write of the *same* location; however, in the case that is allowed, the read data will be discarded (write operation has a high priority).

Table 7.1 shows the realistic FFMs for two-read 2P memories supporting simultaneous read and write of the same location. They are the same as those shown in Table 5.6. However, for (rw-ro)2P memories write operations can only be performed through Prw, while this can be done via any port for (rw-rw)2P memories.

The 2PF1s consist of three FFMs, two require two simultaneous read operations of the same location (i.e., wDRDF&wDRDF and wRDF&wRDF), and one requires simultaneous read and write of the same location (i.e., wRDF&wTF). Since the considered two-read 2P memory allows for such operations, all 2PF1s are realistic.

The 2PF2s consist of three types: $2PF2_a$, $2PF2_v$, and $2PF2_{av}$. The $2PF2_a$s consist of one FFM, which fault primitives (FPs) require simultaneous read and write of the same location (a-cell) or two simultaneous read operations of same a-cell. The $2PF2_v$s consist of two FFMs which require two simultaneous read operations to the v-cell. The $2PF2_{av}$s consist of three FFMs; all of them requires simultaneous read and write operations performed to different locations. Therefore all FFMs of Table 5.6 are realistic for the considered two-read 2P memory.

Table 7.1. Realistic 2PFs for two-read 2P memories; $x \in \{0,1\}$ and $d =$ don't care

Class	FFM	Fault primitives
2PF1	wRDF&wRDF	$< 0r0 : 0r0/1/1 >, < 1r1 : 1r1/0/0 >$
	wDRDF&wDRDF	$< 0r0 : 0r0/1/0 >, < 1r1 : 1r1/0/1 >$
	wRDF&wTF*	$< 0r0 : 0w1/0/- >^*, < 1r1 : 1w0/1/- >^*$
$2PF2_a$	wCFds&wCFds	$< dw0 : drd; 0/1/- >^*, < dw0 : drd; 1/0/- >^*$
		$< dw1 : drd; 0/1/- >^*, < dw1 : drd; 1/0/- >^*$
		$< xrx : xrx; 0/1/- >, < xrx : xrx; 1/0/- >$
$2PF2_v$	wCFrd&wRDF	$< 0; 0r0 : 0r0/1/1 >, < 0; 1r1 : 1r1/0/0 >,$
		$< 1; 0r0 : 0r0/1/1 >, < 1; 1r1 : 1r1/0/0 >$
	wCFdrd&wDRDF	$< 0; 0r0 : 0r0/1/0 >, < 0; 1r1 : 1r1/0/1 >,$
		$< 1; 0r0 : 0r0/1/0 >, < 1; 1r1 : 1r1/0/1 >,$
$2PF2_{av}$	wCFds&wRDF	$< dw0 : 0r0/1/1 >, < dw0 : 1r1/0/0 >,$
		$< dw1 : 0r0/1/1 >, < dw1 : 1r1/0/0 >$
	wCFds&wIRF	$< dw0 : 0r0/0/1 >, < dw0 : 1r1/1/0 >,$
		$< dw1 : 0r0/0/1 >, < dw1 : 1r1/1/0 >$
	wCFds&wRRF	$< dw0 : 0r0/0/? >, < dw0 : 1r1/1/? >,$
		$< dw1 : 0r0/0/? >, < dw1 : 1r1/1/? >$

In the case the design of the two-read 2P memory does not support simultaneous read and write of the same location, the set of the 2PFs can be reduced. The 2PF1: wRDF&wTF will be not realistic; in addition, the $2PF2_a$: $wCF_{ds}\&wCF_{ds}$ will consist only of FPs sensitized by two simultaneous read operations. The net result can be obtained from Table 7.1 by removing the FFMs and the FPs marked with '*'.

7.2.2 Realistic 2PFs for single-read 2P memories

The single-read 2P memories consists of 2P memories that do not support two simultaneous read operations. Therefore, all FFMs sensitized with this kind of operations are not realistic. In addition, and depending on their design, such memories can allow/not allow for simultaneous read and write of the *same* location (i.e., 'wx_c:ry_c', $x, y \in \{0, 1\}$).

Table 7.2 show the realistic FFMs for single-read 2P memories supporting 'wx_c:ry_c'; see also Table 7.1. Since no simultaneous read operations to the same location are supported, the 2PF1: wDRDF&wDRDF and wRDF&wRDF, as well as 2PF2$_v$ are not realistic; this is because they require two simultaneous operations for their sensitization. On the other hand, the 2PF1: wRDF&wTF as well as the 2PF2$_a$ will be realistic; however, the 2PF2$_a$ will consist only of FPs based on simultaneous read and write. The 2PF2$_{av}$ is realistic since it requires simultaneous read and write of *different* locations, which is supported by the single-read 2P memories.

Table 7.2. Realistic 2PFs for single-read 2P memory supporting 'wx_c:ry_c'

Class	FFM	Fault primitives
2PF1	wRDF&wTF	$< 0r0 : 0w1/0/- >, < 1r1 : 1w0/1/- >$
2PF2$_a$	wCFds&wCFds	$< dw0 : drd; 0/1/- >, < dw0 : drd; 1/0/- >$ $< dw1 : drd; 0/1/- >, < dw1 : drd; 1/0/- >$
2PF2$_v$	No fault	-
2PF2$_{av}$	wCFds&wRDF	$< dw0 : 0r0/1/1 >, < dw0 : 1r1/0/0 >,$ $< dw1 : 0r0/1/1 >, < dw1 : 1r1/0/0 >$
	wCFds&wIRF	$< dw0 : 0r0/0/1 >, < dw0 : 1r1/1/0 >,$ $< dw1 : 0r0/0/1 >, < dw1 : 1r1/1/0 >$
	wCFds&wRRF	$< dw0 : 0r0/0/? >, < dw0 : 1r1/1/? >,$ $< dw1 : 0r0/0/? >, < dw1 : 1r1/1/? >$

In the case 'wx_c:ry_c' is not supported by the memory, all faults in Table 7.2 sensitized by simultaneous read and write operation to the same location have to be considered not realistic. These are the 2PF1: wRDF&wTF, and the 2PF2$_a$. The result of this simplification is shown in Table 7.3. Note that the set of realistic FFMs for such memories consists only of 2PF2$_{av}$.

7.3 Tests for restricted two-port memories

The 2P tests introduced in Chapter 6 are valid for (rw-rw)2P memories. It has been shown in the previous section that the port restrictions impact the realistic set of FFMs; and hence they also impact the tests. In the rest of this section, first the tests for two-read 2P memories will be established, thereafter for single-read 2P memories.

Table 7.3. Realistic 2PFs for single-read 2P memory not supporting 'wx_c:ry_c'

Class	FFM	Fault primitives
2PF1	No fault	-
2PF2$_a$	No fault	-
2PF2$_v$	No fault	-
2PF2$_{av}$	wCFds&wRDF	$< dw0 : 0r0/1/1 >$, $< dw0 : 1r1/0/0 >$, $< dw1 : 0r0/1/1 >$, $< dw1 : 1r1/0/0 >$
	wCFds&wIRF	$< dw0 : 0r0/0/1 >$, $< dw0 : 1r1/1/0 >$, $< dw1 : 0r0/0/1 >$, $< dw1 : 1r1/1/0 >$
	wCFds&wRRF	$< dw0 : 0r0/0/? >$, $< dw0 : 1r1/1/? >$, $< dw1 : 0r0/0/? >$, $< dw1 : 1r1/1/? >$

7.3.1 Tests for two-read two-port memories

It has been shown in Section 7.2.1 that 2PFs for two-read 2P memories consist of the same faults as those introduced in Chapter 5, for which a set of tests has been developed in Chapter 6 and summarized in Table 6.6. Therefore, the same tests are applicable to this class of 2P memories; except for the second version of March d2PF- (see Figure 6.13). This test requires that, in addition to the read capability of both ports, the ports should also have the write capability. Therefore, that test is only applicable to (rw-rw)2P memories, and not to (rw-ro)2P memories. Tests for (rw-ro)2P memories have to be applied in such way that all write operations have to be performed via P_a=Prw.

7.3.2 Tests for single-read two-port memories

The single-read 2P memories allow at the most for one read operation at a time, and consist of (wo-rw)2P and (wo-ro)2P memories. The 2PFs of such class are shown in Table 7.2 and in Table 7.3, for the case the 2P memory allows, respectively not allows, for simultaneous write and read of the same location.

In the case the 2P memory does not support simultaneous write and read of the same location, then the 2PFs consist only of the 2PF2$_{av}$ faults, and therefore only March d2PF2 (see Figure 6.11), or its optimized version of Figure 6.12, is required. For the two types, (wo-rw)2P and (wo-ro)2P memories, the write operations have to be applied via P_a and the read operations via P_b.

In the case the singe-read 2P memories support simultaneous write and read of the same location, the 2PFs consist not only of the 2PF2$_{av}$ but also of the 2PF1, wRDF&wTF, and of the 2PF2$_a$; however, the latter consists only of FPs based on simultaneous write and read operations (see Table 7.2). Therefore, in addition to March d2PF2 (which detects the 2PF2$_{av}$), two tests should be specified in order to detect the 2PF1 and 2PF2$_a$ faults.

Test for wRDF&wTF

The test detecting the wRDF&wTF in single-read 2P memories is shown in Figure
7.1; it has a test length of $5n$ and satisfies Condition wRDF&wTF of Section 6.3.1
(Chapter 6). M_1 contains the pairs of march elements of Case A merged into a single
march element; while M_2 contains the pairs of march elements of Case B merged into
a single march element. Note that in the test, all write operations are performed
via P_a, and read operations via P_b, such that the test is compatible with the type of
ports the (wo-rw)2P and (wo-ro)2P memories consist of.

$$\{ \, \updownarrow (w0:n) \; ; \; \updownarrow (w1:r0,n:r1) \quad \updownarrow (w0:r1,r1) \, \}$$
$$M_0 \qquad\qquad\qquad M_1 \qquad\qquad\qquad M_2$$

Figure 7.1. March 2PF1 for single-read 2P memories

Test for 2PF2$_a$

The 2PF2$_a$ for single-read 2P memories consists of one FFM: wCF$_{ds}$&wCF$_{ds}$; with
FPs based only on simultaneous read and write to the same location; see Table 7.2.
An optimal test detecting such faults is shown in Figure 7.2 and has a test length of
$10n$; it is a modified version of March 2PF2$_a$- of Figure 6.8. The sensitizing/detection
operations based on two simultaneous read operations, which are not supported in
single-read 2P memories, are changed into single read operations. In the test of
Figure 7.2, all write operations are applied via P_a, and all read operations via P_b,
such that the test can be applied to (wo-rw)2P and (wo-ro)2P memories. The test
of Figure 7.2 detects all 2PF2$_a$s in single-read 2P memories (see Table 7.2):

- The $< dw1 : drd; 0/1/- >_{a,v}$ will be sensitized by the operation '$w1 : r0$' and
  detected by '$n : r0$' in M_1 of Figure 7.2 if the v-cell has a higher address than
  the a-cell; i.e., $v > a$. If $v < a$, then the fault will be sensitized and detected
  by M_3.

$$\{ \, \updownarrow (w0:n) \; ; \; \Uparrow (n:r0,w1:r0) \; ; \; \Uparrow (n:r1,w0:r1) \quad$$
$$M_0 \qquad\qquad M_1 \qquad\qquad\qquad M_2$$
$$\Downarrow (n:r0,w1:r0) \; ; \; \Downarrow (n:r1,w0:r1) \; ; \; \Downarrow (n:r0) \, \}$$
$$M_3 \qquad\qquad\qquad M_4 \qquad\qquad M_5$$

Figure 7.2. March 2PF2$_a$- for single-read 2P memories

- The $< dw0 : drd; 1/0/- >_{a,v}$ will be sensitized and detected by M_2 if $v > a$. If
  $v < a$, then the fault will be sensitized and detected by M_4.

- The $< dw1 : drd; 1/0/- >_{a,v}$ will be sensitized by M_1 and detected by M_2 if
  $v < a$; while the same fault will be sensitized by M_3 and detected by M_4 if
  $v > a$.

- The $< dw0 : drd; 0/1/- >_{a,v}$ will be sensitized by M_2 and detected by M_3 if $v < a$; while the same fault will be sensitized by M_4 and detected by M_5 if $v > a$.

It can be verified easily that March $2PF2_a$ also detects the two FPs of wRDF&wTF:

- The $< 0r0 : 0w1/0/- >$ is sensitized by '$w1 : r0$" in 'M_1 (also by M_3) and detected by '$r1 : n$' in M_2 (also by M_4).

- The $< 1r1 : 1w0/1/- >$ is sensitized by M_2 (also by M_4) and detected by M_3 (also by M_5).

Therefore one can use only March $2PF2_a$- to detect all realistic 2PFs in single-read 2P memories. From now on, this march test will be referred to as *March s2PF-* for single-read 2P memories; such that testing of the 2PFs in single-read 2P memories (supporting simultaneous read and write to the same location) can be done with that march test and March d2PF-.

7.4 Test strategy for restricted two-port memories

Applying march tests requires performing reads as well as write operations. Therefore, it is not always possible to apply a test via any port of the memory; for instance, a march test can not be applied via Pro, neither via Pwo. That means that the type of ports the memory consists of has a great impact on the test strategy. In the rest of this section, the test strategy for the two-read and single-read 2P memories will be discussed.

7.4.1 Test strategy for two-read 2P memories

The two-read 2P memories consist of (rw-rw)2P and (rw-wo)2P memories. The test strategy for (rw-rw)2P has been discussed in Section 6.6; it is repeated in Figure 7.3. In order to appreciate the test time consequence for the test strategy, and compare it with the time required for other types of 2P memories, it will be assumed that only March SS is used for testing 1PFs. It has been shown in Section 6.6 that the test procedure in that case will require a test time of $67n$.

A similar test strategy can be applied to (rw-ro)2P memories; see Figure 7.4. However, in that case, all write operations should be applied via Prw; this applies for SP tests as well as for 2P tests. Note that the total test time required for (rw-ro)2P memories is the same as that for (rw-rw)2P memories, since March SS is performed two times, March s2PF- once, and March d2PF- once. Note that for testing (rw-ro)2P memories, only the the test of Figure 6.12 can be used to detect 2PF2s; *not* the test of Figure 6.13 as is the case for (rw-rw)2P memories; see Figure 7.3. As mentioned in Section 7.3.1, the test of Figure 6.13 requires that, in addition to the read capability, both ports should have also the write capability.

```
Detection of 1PFs:
    Apply single port test(s) (e.g., March SS) via port Pₐ;
    If die fails, go to FAIL;
    Apply single port test(s) (e.g., March SS) via port P_b;
Detection of 2PFs:
    If die fails, go to FAIL;
    Apply March s2PF- of Figure 6.15 via the two ports;
    If die fails, go to FAIL;
    Apply March d2PF- of Figure 6.12 or of Figure 6.13 via the two ports;
    If die fails, go to FAIL;
Pass: Print 'Die passes';
      END;
Fail: Print 'Die fails';
      END;
```

Figure 7.3. Test strategy for (rw-rw)2P memories

```
Detection of 1PFs:
    Apply, e.g., March SS via port Prw;
    If die fails, go to FAIL;
    Apply e.g., March SS via port Pro in such way that:
        Write operations will be performed via Prw;
        and read operations will be performed via Pro;
    If die fails, go to FAIL;
Detection of 2PFs:
    Apply March s2PF- of Figure 6.15 via the two ports;
    If die fails, go to FAIL;
    Apply March d2PF- of Figure 6.12 via the two ports;
    If die fails, go to FAIL;
Pass: Print 'Die passes';
      END;
Fail: Print 'Die fails';
      END;
```

Figure 7.4. Test strategy for (rw-ro)2P memories

7.4.2 Test strategy for single-read 2P memories

The single-read 2P memories consist of (wo-rw)2P and (wo-ro)2P memories. Their realistic FFMs and required tests depend on their support of simultaneous write and read of the same location (i.e., '$wx_c{:}ry_c$', $x, y \in \{0, 1\}$); see Table 7.2 and Table 7.3.

Test strategy for memories not supporting 'wx_c:ry_c'

The test strategy for (wo-rw)2P memories is shown in Figure 7.5. To detect the 1PFs, March SS is first applied via Prw; thereafter via Pwo in such way that the write operations are applied via Pwo, while the read operations are performed via Prw. Since the 2PFs for such memories consist only of $2PF2_{av}$, only March $2PF2_{av}$- of Figure 6.12 needs to be applied. Note that the total test time required is $53n$, since March SS with a test length $22n$ is performed twice; and march $2PF2_{av}$-, with a test length of $9n$, is performed once.

```
Detection of 1PFs:
     Apply, e.g., March SS via port Prw;
     If die fails, go to FAIL;
     Apply e.g., March SS via port Pwo in such way that:
        Write operations will be performed via Pwo;
        and read operations will be performed via Prw;
     If die fails, go to FAIL;
Detection of 2PFs:
        Apply March d2PF- of Figure 6.12 via the two ports;
     If die fails, go to FAIL;
Pass: Print 'Die passes';
        END;
Fail: Print 'Die fails';
        END;
```

Figure 7.5. Test strategy for (wo-rw)2P memories not supporting 'wx_c:ry_c'

```
Detection of 1PFs:
     Apply e.g., March SS in such way that:
        Write operations will be performed via Pwo;
        and read operations will be performed via Pro;
        If die fails, go to FAIL;
Detection of 2PFs:
        Apply March d2PF- of Figure 6.12 via the two ports;
        If die fails, go to FAIL;
Pass: Print 'Die passes';
        END;
Fail: Print 'Die fails';
        END;
```

Figure 7.6. Test strategy for (wo-ro)2P memories not supporting 'wx_c:ry_c'

A similar test strategy can be applied to (wo-ro)2P memories; see Figure 7.6. Note that no SP test can be applied via Pwo, nor via Pro. The only possibility is to apply the test in such way that the write operations will be done via Pwo, and the read operations via Pro. To detect the 2PFs, March $2PF2_{av}$- has to be performed through the two ports. Note that the total test time required is $31n$, since both March SS and March $2PF2_{av}$ are performed each once.

Test strategy for memories supporting 'wx_c:ry_c'

The test strategy for (wo-rw)2P memories is given in Figure 7.7, and is similar to that of Figure 7.5. The only difference is that an extra test, March s2PF- of Figure 7.2, should be applied in order to detect the 2PF1: wRDF&wTF, and the $2PF2_a$ faults. The Total test time becomes than $63n$.

Detection of 1PFs:
 Apply, e.g., March SS via port Prw;
 If die fails, go to FAIL;
 Apply e.g., March SS via port Pwo in such way that:
   Write operations will be performed via Pwo;
   and read operations will be performed via Prw;
 If die fails, go to FAIL;
Detection of 2PFs:
 Apply March s2PF- of Figure 7.2 via the two ports;
 If die fails, go to FAIL;
 Apply March d2PF- of Figure 6.12 via the two ports;
 If die fails, go to FAIL;
Pass: Print 'Die passes';
    END;
Fail: Print 'Die fails';
    END;

Figure 7.7. Test strategy for (wo-rw)2P memories supporting 'wx_c:ry_c'

Figure 7.8 shows the test strategy for (wo-ro)2P memories; it is the same as that of Figure 7.6; except it is extended with an extra test to detect the wRDF&wTF, and the $2PF2_a$ faults. The Total test time becomes than $41n$.

Detection of 1PFs:
  Apply e.g., March SS in such way that:
    Write operations will be performed via Pwo;
    and read operations will be performed via Pro;
  If die fails, go to FAIL;
Detection of 2PFs:
  Apply March s2PF- of Figure 7.2 via the two ports;
  If die fails, go to FAIL;
  Apply March d2PF- of Figure 6.12 via the two ports;
  If die fails, go to FAIL;
Pass: Print 'Die passes';
      END;
Fail: Print 'Die fails';
      END;

Figure 7.8. Test strategy for (wo-ro)2P memories supporting 'wx_c:ry_c'

Part III

Testing p-port SRAMs

Chapter 8. Experimental analysis of p-port SRAMs

Chapter 9. Tests for p-port SRAMs

Chapter 10. Testing restricted p-port SRAMs

This part develops realistic fault models, tests, and a test strategy for any multi-port memory with p ports (p > 2). It is divided into three chapters. Chapter 8 introduces realistic fault models, together with their probabilities, based on defect injection and SPICE simulation for three-port memories; thereafter the results are extended for any multi-port memory. Chapter 9 establishes tests for the introduced faults. Chapter 7 covers the impact of port restriction (e.g., read-only port, write-only port) on the faults models, tests, as well as on the test strategy.

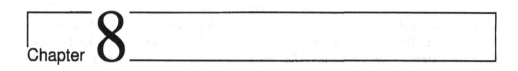

Chapter **8**

Experimental analysis of p-port SRAMs

In Chapter 5 through Chapter 7, realistic fault models, based on defect injection and circuit simulation, together with their tests have been presented for two-port memories. In addition, the impact of port restrictions on the fault models, as well as on the tests, have been addressed.

In this chapter, the simulation will be done for three-port memories, in order to derive realistic fault models. The results will thereafter be extended for any multi-port memory with p ports. Section 8.1 derives the to-be-simulated spot defects, based on the analysis done in Chapter 4 for a p-port memory. Section 8.2 presents the electrical simulation results using the fault primitive notation. Section 8.3 transforms the electrical faults into realistic functional fault models. Section 8.4 gives the probability of occurrence of such faults by assuming that all modeled spot defects are equal likely to occur. Section 8.5 derives the fault models for any p-port memory, based on the work done for three-port memories, and for two-port memories in Chapter 5.

8.1   The to-be simulated spot defects

A complete analysis of spot defects (SDs) (which have been modeled as opens, shorts, and bridges) in a differential multi-port (MP) memory cell has been done in Chapter 4. In Chapter 5, the set of SDs for two-port (2P) memories has been derived. This set has been simulated in order to inspect the behavior of the memory in the presence of any defect. Based on those simulation results, and depending on the kind of the faults sensitized in their presence, SDs can be classified into *single-port fault defects (SFDs)* and *multi-port fault defects (MFDs)*. The SFDs are defects that can *only* cause single-port (SP) faults; they can not cause faults typical for MP memories. However, MFDs are defects that can cause SP faults as well as faults typical for MP memories (e.g., 2P faults). That means that in order to analyze any MP memory with $p > 2$, one needs to concentrate only on MFDs since faults that can be caused with SFDs are SP faults; they do not depend on the number of ports of the MP memory.

A differential three-port (3P) memory cell, with three read-write ports, will be the subject of this chapter; see Figure 4.1 with $p = 3$. It will be assumed that the 3P memory allows for the same SP and 2P operations as those for the 2P memory considered in Section 5.1. In addition, it will be assumed that the 3P memory supports the following 3P operations:

- Three simultaneous read operations to the same location.

- Three simultaneous operations (read and/or write) operations to three different locations.

- Two simultaneous read operations to one location (cell c_1) and another operation (read or write) to another location, c_2, at a time (i.e., '$rx_{c1} : rx_{c1} : ry_{c2}$' or '$rx_{c1} : rx_{c1} : wy_{c2}$'; $x, y \in \{0, 1\}$).

- A simultaneous read and a write operation to one location, c_1, and another operation (read or write) to another location, c_2, at a time (i.e., '$rx_{c1} : wy_{c1} : rz_{c2}$' or '$rx_{c1} : wy_{c1} : wz_{c2}$'; $x, y, z \in \{0, 1\}$).

- Two simultaneous reads and write to the same location (i.e., '$rx_{c1} : rx_{c1} : wy_{c1}$'). However, in that case the read data will be discarded; i.e., the write operation has a high priority.

In Chapter 4, all opens, shorts and bridges have been defined and located for a differential p-port memory; $p > 1$. For $p = 3$, it can be verified easily that there are 223 possible SDs: 53 opens, 22 shorts, and 148 bridges. However, the 223 possible SDs in a 3P memory cell can be placed into 49 groups, whereby only one SD from each group needs to be simulated; the behavior of other SDs within a group can be derived based on the simulated one. The grouping is based on the fact that the memory cell has a symmetrical structure and with similar ports; see Section 4.2. Moreover, only MFDs will be considered; see Table 4.1 through Table 4.4. The minimal set of MFDs that has to be simulated for a 3P memory consists of 18 SDs: 2 opens, 1 short, and 15 bridges; they are given in Table 8.1 and Table 8.2. Note that the bridges are divided into bridges within a cell (BCs), and bridges between cells (BCCs). The BCCs are further divided into BCCs in the same row (rBCCs), in the same column (cBCCs), and on the same diagonal (dBCCs); see Figure 4.4.

Table 8.1. Minimal set of opens and shorts to-be simulated for a 3P memory

MFD	Description
OC3	Drain of pull-down at true (false) side broken
OC4	Source of pull-down at true (false) side broken
SC2	True (false) node shorted to V_{ss}

Table 8.2. Minimal set of bridges to-be simulated for a 3P memory

Bridges	BCs	Bridges	rBCCs	Bridges	cBCCs	Bridges	dBCCs
BC6	BL_a-BL_b	rBCC1	T1-T3	cBCC1	T1-T2	dBCC1	T1-T4
BC7	BL_a-$\overline{BL_b}$	rBCC2	T1-F3	cBCC2	T1-F2	dBCC2	T1-F4
		rBCC3	T1-$BL2_a$	cBCC3	T1-$WL2_a$		
		rBCC4	T1-$\overline{BL2_a}$				
		rBCC5	$BL1_a - BL2_a$				
		rBCC6	$BL1_a - BL2_b$				
		rBCC7	$BL1_a - \overline{BL2_a}$				
		rBCC8	$BL1_a - \overline{BL2_b}$				

8.2 Simulation results

The simulation has been done for all 18 SDs listed in Table 8.1 and Table 8.2, by using the simulation methodology of Chapter 4. Each electrical faulty behavior is reported in terms of a fault primitive (FP); if a strong fault is sensitized, then the FP notations introduced in Chapter 3 are used to describe it. The notation is extended to denote *three-port faults (3PFs)* (i.e., faults requiring the use of the three ports simultaneously in order to be sensitized); for example: '$< 0r0 : 0r0 : 0r0/1/1 >_v$' denotes a FP sensitized in the v-cell by applying three simultaneous read operations to the v-cell; the v-cell then flips, and the sense amplifiers return incorrect values. If a fault is only partially sensitized (e.g., weak fault) then the fault is denoted as wF.

In order to save space, only simulation results for some MFDs will be presented here; the results of all simulated MFDs are given in Appendix B. Table 8.3 lists the

Table 8.3. Overview of the simulation results for some MFDs; $d = $ *don't care value*

Label	R_{df}	Fault primitive	Compl. fault primitive	Class	Fault model
OC3	I	wF	wF	-	-
	II	$< 0r0 : 0r0 : 0r0/1/0 >_v$	$< 1r1 : 1r1 : 1r1/0/1 >_v$	3PF1	wDRDF&wDRDF&wDRDF
	III	$< 0r0 : 0r0 : 0r0/1/1 >_v$	$< 1r1 : 1r1 : 1r1/0/0 >_v$	3PF1	wRDF&wRDF&wRDF
	IV	$< 0r0 : 0r0/1/0 >_v$	$< 1r1 : 1r1/0/1 >_v$	2PF1	wDRDF&wDRDF
	V	$< 0r0 : 0r0/1/1 >_v$	$< 1r1 : 1r1/0/0 >_v$	2PF1	wRDF&wRDF
	VI	$< 0r0/1/0 >_v$	$< 1r1/0/1 >_v$	1PF1	DRDF
	VII	$< 0r0/1/1 >_v$	$< 1r1/0/0 >_v$	1PF1	RDF
	VIII	$< 0r0/1/1 >_v$	$< 1r1/0/0 >_v$	1PF1	RDF
		$< 0_T/1/- >_v$	$< 1_T/0/- >_v$	1PF1	DRF
SC2	I	$< 0/1/- >_v$	$< 1/0/- >_v$	1PF1	SF
	II	$< 1r1/0/0 >_v$	$< 0r0/1/1 >_v$	1PF1	RDF
	III	$< 1r1/0/1 >_v$	$< 0r0/1/0 >_v$	1PF1	DRDF
	IV	$< 1r1 : 1r1/0/0 >_v$	$< 0r0 : 0r0/1/1 >_v$	2PF1	wRDF&wRDF
	V	$< 1r1 : 1r1/0/1 >_v$	$< 0r0 : 0r0/1/0 >_v$	2PF1	wDRDF&wDRDF
	VI	$< 1r1 : 1r1 : 1r1/0/0 >_v$	$< 0r0 : 0r0 : 0r0/1/1 >_v$	3PF1	wRDF&wRDF&wRDF
	VII	$< 1r1 : 1r1 : 1r1/0/1 >_v$	$< 0r0 : 0r0 : 0r0/1/0 >_v$	3PF1	wDRDF&wDRDF&wDRDF
	VIII	wF	wF	-	-
BC6	I	$< dw0 : 1r1/0/0 >_{a,v}$	$< dw1 : 0r0/1/1 >_{a,v}$	2PF2	wCFds&wRDF
	II	$< dw0 : 1r1/1/0 >_{a,v}$	$< dw1 : 0r0/0/1 >_{a,v}$	2PF2	wCFds&wIRF
	III	$< dw0 : 1r1/1/? >_{a,v}$	$< dw1 : 0r0/0/? >_{a,v}$	2PF2	wCFds&wRRF
	IV	wF	wF	-	-
BC7	I	$< 0r0 : 0w1/0/ >_v$		2PF1	wRDF&wTF
		$< 1r1 : 1w0/1/ >_v$		2PF1	wRDF&wTF
		$< w1 : 1r1/0/0 >_{a,v}$		2PF2	wCFds&wRDF
		$< dw0 : 0r0/1/1 >_{a,v}$		2PF2	wCFds&wRDF
	II	$< dw1 : 1r1/0/0 >_{a,v}$		2PF2	wCFds&wRDF
		$< dw0 : 0r0/1/1 >_{a,v}$		2PF2	wCFds&wRDF
	III	$< dw1 : 1r1/1/0 >_{a,v}$		2PF2	wCFds&wIRF
		$< dw0 : 0r0/0/1 >_{a,v}$		2PF2	wCFds&wIRF
	IV	$< dw1 : 1r1/1/? >_{a,v}$		2PF2	wCFds&wRRF
		$< dw0 : 0r0/0/? >_{a,v}$		2PF2	wCFds&wRRF
	V	wF		-	-
rBCC1	I	$< 0; 1/0/- >_{a,v}$	$< 1; 0/1/- >_{a,v}$	1PF2	CFst
	II	$< 0; 1r1/0/0 >_{a,v}$	$< 1; 0r0/1/1 >_{a,v}$	1PF2	CFrd
		$< 0r0; 1/0/- >_{a,v}$	$< 1r1; 0/1/- >_{a,v}$	1PF2	CFds
	III	$< 0; 1r1 : 1r1/0/0 >_{a,v}$	$< 1; 0r0 : 0r0/1/1 >_{a,v}$	2PF2	wCFrd&wRDF
		$< 0r0 : 0r0; 1/0/- >_{a,v}$	$< 1r1 : 1r1; 0/1/- >_{a,v}$	2PF2	wCFds&wCFds
		$< dw0 : drd; 1/0/- >_{a,v}$	$< dw1 : drd; 0/1/- >_{a,v}$	2PF2	wCFds&wCFds
	IV	$< 0; 1r1 : 1r1/0/1 >_{a,v}$	$< 1; 0r0 : 0r0/1/0 >_{a,v}$	2PF2	wCFdrd&wDRDF
		$< 0r0 : 0r0; 1/0/- >_{a,v}$	$< 1r1 : 1r1; 0/1/- >_{a,v}$	2PF2	wCFds&wCFds
		$< dw0 : drd; 1/0/- >_{a,v}$	$< dw1 : drd; 0/1/- >_{a,v}$	2PF2	wCFds&wCFds
	V	$< 0; 1r1 : 1r1 : 1r1/0/0 >_{a,v}$	$< 1; 0r0 : 0r0 : 0r0/1/1 >_{a,v}$	3PF2	wCFrd&wRDF&wRDF
		$< 0r0 : 0r0 : 0r0; 1/0/- >_{a,v}$	$< 1r1 : 1r1 : 1r1; 0/1/- >_{a,v}$	3PF2	wCFds&wCFds&wCFds
		$< dw0 : drd : drd; 1/0/- >_{a,v}$	$< dw1 : drd : drd; 0/1/- >_{a,v}$	3PF2	wCFds&wCFds&wCFds
	VI	$< 0; 1r1 : 1r1 : 1r1/0/1 >_{a,v}$	$< 1; 0r0 : 0r0 : 0r0/1/0 >_{a,v}$	3PF2	wCFdrd&wDRDF&wDRDF
		$< 0r0 : 0r0 : 0r0; 1/0/- >_{a,v}$	$< 1r1 : 1r1 : 1r1; 0/1/- >_{a,v}$	2PF2	wCFds&wCFds&wCFds
		$< dw0 : drd : drd; 1/0/- >_{a,v}$	$< dw1 : drd : drd; 0/1/- >_{a,v}$	3PF2	wCFds&wCFds&wCFds
	VII	wF	wF	-	-
cBCC1	I	$< 0; 1/0/- >_{a,v}$	$< 1; 0/1/- >_{a,v}$	1PF2	CFst
	II	$< 0; 1r1/0/0 >_{a,v}$	$< 1; 0r0/1/1 >_{a,v}$	1PF2	CFrd
	III	$< 0; 1r1/0/1 >_{a,v}$	$< 1; 0r0/1/0 >_{a,v}$	1PF2	CFdrd
	IV	$< 0; 1r1 : 1r1/0/0 >_{a,v}$	$< 1; 0r0 : 0r0/1/1 >_{a,v}$	2PF2	wCFrd&wRDF
	V	$< 0; 1r1 : 1r1/0/1 >_{a,v}$	$< 1; 0r0 : 0r0/1/0 >_{a,v}$	2PF2	wCFdrd&wDRDF
	VI	$< 0; 1r1 : 1r1 : 1r1/0/0 >_{a,v}$	$< 1; 0r0 : 0r0 : 0r0/1/1 >_{a,v}$	3PF2	wCFrd&wRDF&wRDF
	VII	$< 0; 1r1 : 1r1 : 1r1/0/1 >_{a,v}$	$< 1; 0r0 : 0r0 : 0r0/1/0 >_{a,v}$	3PF2	wCFdrd&wDRDF&wDRDF
	VIII	wF	wF	-	-

simulation results for one open, one short, two bridges within a cell, and two bridges between cells. The first column in the table gives the name of the simulated SD; see also Table 8.1 and Table 8.2. The second column gives the resistance regions[1] ordered

---

[1]The exact resistance values for each region are process specific and Intel proprietary

in an increasing values, the third and the fourth columns list the FP sensitized by the simulated SD and the derived complementary FP (if applicable), respectively; a complementary FP is applicable if there is another modeled SD which has a complementary fault behavior as that of the SD simulated (see Section 4.3). The fifth column gives the class of the sensitized fault; i.e., single-port faults involving a single cell (1PF1s), single-port faults involving two cells (1PF2s), two-port faults involving a single cell (2PF1s), two-port faults involving two cells (2PF2s), three-port faults involving a single-cell (3PF1s), and three-port faults involving two cells (3PF2s). The sixth column shows the FFM each FP belongs to. Note that for the description of the FFMs for 3PFs, a similar notation as that of 2PF FFMs is used; e.g., the FFM 'wRDF&wRDF&wRDF' denotes a 3PF based on three weak Read Destructive Faults. Table 8.3 clearly shows that the sensitized faults are strongly dependent on the resistance value of the SD. It is also interesting to note that the bridges BC6 and BC7 (i.e., bridges between bit lines belonging to the same column and different ports) can only cause 2PF1s and 2PF2s; they don't cause any 3PFs.

8.3 Realistic fault models for three-port memories

The electrical faults, expressed in terms of FPs, caused by MFDs (i.e., opens, shorts, and bridges) as well as those caused by SFDs (see Chapter 5) can be translated into sets of *realistic* FFMs for 3P memories; remember that a FFM is defined as a non-empty set of FPs. Based on the number of ports required in order to sensitize a fault, the realistic FFMs for memory cell array faults in 3P memories can be classified into three classes [25, 31]; see Figure 8.1.

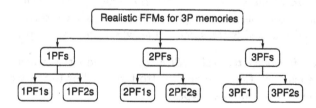

Figure 8.1. Classification of realistic FFMs in 3P memories

1. Single-port faults (1PFs): These are faults requiring the use of, at the most, one port in order to be sensitized. They are divided into 1PFs involving a single cell (1PF1s), and 1PFs involving two-cells (1PF2s). The 1PFs have been discussed in details in Section 5.3.1.

2. Two-port faults (2PFs): These are faults that can not be sensitized using SP operations; they require the use of two ports simultaneously. They are also

divided into 2PFs involving a single cell (2PF1s) and 2PFs involving two cells (2PF2s). The 2PFs are addressed in detail in Section 5.3.2.

3. Three-port faults (3PFs): These are faults that can not be sensitized using SP operations, neither using 2P operations; they require the use of the three ports of the memory simultaneously. The 3PFs can be also divided into 3PFs involving a single cell (3PF1s) and 3PFs involving two cells (3PF2s); they are discussed in detail in the following subsections.

8.3.1   Realistic three-port faults

As is mentioned above, three-port faults (3PFs) can not be sensitized using SP operations, neither by 2P operations; they require the use of the three ports of the memory *simultaneously*. They are divided into two subclasses; see Figure 8.2.

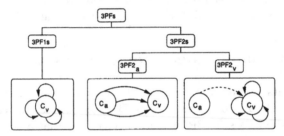

Figure 8.2. Classification of 3PFs

1. *The 3PFs involving a single cell (3PF1s)*: They are based on a combination of three single port operations applied simultaneously via three ports to a single cell; the cell accessed is the same cell as where the fault appears.

2. *The 3PFs involving two cells (3PF2s)*: Depending to which cells the three simultaneous operations are applied (to the a-cell or to the v-cell), the 3PF2s are further divided into two types:

   1. The $3PF2_a$: this fault is sensitized in the v-cell c_v by applying three simultaneous operations to the a-cell c_a.

   2. The $3PF2_v$: this fault is sensitized in the v-cell by applying three simultaneous operations to the v-cell (solid arrows in Figure 8.2), while the a-cell is in a certain state (dashed arrow in the figure).

Table 8.4 lists the realistic FFMs for 3PFs, together with the FPs they consists of; it contains the entities of Table 8.3 with 3PF1 and 3PF2 in the column 'Class'. These FFMs are listed below together with defects causing them. Note that all FFMs for 3PFs are a combination of three weak faults; remember that a weak fault can not fully sensitize a fault; however, the fault effects of two or more weak faults may be additive and hence fully sensitize a fault when the weak faults are sensitized simultaneously.

Table 8.4. List of realistic 3PF FFMs; $x \in \{0, 1\}$ and $d =$ don't care

Class	#	FFM	Fault primitives
3PF1	1	wRDF&wRDF&wRDF	$< 0r0 : 0r0 : 0r0/1/1 >$, $< 1r1 : 1r1 : 1r1/0/0 >$
	2	wDRDF&wDRDF&wDRDF	$< 0r0 : 0r0 : 0r0/1/0 >$, $< 1r1 : 1r1 : 1r1/0/1 >$
3PF2$_a$	3	wCFds&wCFds&wCFds	$< dw0 : drd : drd; 0/1/- >$, $< dw0 : drd : drd; 1/0/- >$,
			$< dw1 : drd : drd; 0/1/- >$, $< dw1 : drd : drd; 1/0/- >$,
			$< xrx : xrx : xrx; 0/1/- >$, $< xrx : xrx : xrx; 1/0/- >$
3PF2$_v$	4	wCFrd&wRDF&wRDF	$< 0; 0r0 : 0r0 : 0r0/1/1 >$, $< 0; 1r1 : 1r1 : 1r1/0/0 >$,
			$< 1; 0r0 : 0r0 : 0r0/1/1 >$, $< 1; 1r1 : 1r1 : 1r1/0/0 >$
	5	wCFdrd&wDRDF&wDRDF	$< 0; 0r0 : 0r0 : 0r0/1/0 >$, $< 0; 1r1 : 1r1 : 1r1/0/1 >$,
			$< 1; 0r0 : 0r0 : 0r0/1/0 >$, $< 1; 1r1 : 1r1 : 1r1/0/1 >$

Realistic 3PF1 faults

Based on the simulation results of the MFDs, two realistic FFMs have been established (see also the sixth column of Table 8.3). They can be considered as a combination of three single-cell weak faults.

1. wRDF&wRDF&wRDF: Applying three simultaneous read operations to the v-cell causes the v-cell to flip; while the sense amplifiers return *incorrect values*. In order to detect this fault, at least an extra single read operation has to be performed to the same cell. The wRDF&wRDF&wRDF consists of two FPs: $< 0r0 : 0r0 : 0r0/1/1 >_v$ (i.e., applying three simultaneous $r0$ operations to cell c_v will flip the cell to 1, and the sense amplifiers returns 1), and $< 1r1 : 1r1 : 1r1/0/0 >_v$; it can be caused with the following defects (see Appendix B):

   - Drain of the pull-down transistor of the cell broken (OC3).
   - Source of the pull-down transistor of the cell broken (OC4).
   - True or false node shorted to V_{ss} (SC2).
   - Short between a node of a cell and a word line of an adjacent cell in the same column (cBCC3).

2. wDRDF&wDRDF&wDRDF: Applying three simultaneously read operations to the v-cell causes the v-cell to flip and the sense amplifiers return *correct* values. This is because the flipping of the cell happens relatively slowly. This FFM consists of two FPs: $< 0r0 : 0r0 : 0r0/1/0 >_v$ and $< 1r1 : 1r1 : 1r1/0/1 >_v$; it can be caused by the same defects as those causing wRDF&wRDF&wRDF, but with different resistance values of the defect.

Realistic 3PF2 faults

The 3PF2s, which can be considered as a combination of single-cell weak faults and weak faults involving two cells, are divided into two types; see Figure 8.2. Based on the FPs found by simulation, the following 2PF3s have been derived.

The 3PF2$_a$: This type consists of only one FFM: wCFds&wCFds&wCFds. Applying three simultaneous operations to the a-cell will sensitize a fault in the v-cell; i.e., the v-cell flips. This FFM consists of eight FPs; see Table 8.4. Note that, e.g., '< $dw0 : drd : drd; 0/1/-$ >' denotes one FP since the read data is irrelevant, and that the '< $xrx : xrx : xrx; 1/0/-$ >' denotes two FPs since x can be 0 or 1. The 3PF2$_a$s can be caused by (a) Bridges between nodes of adjacent cells belonging to the same row (rBCC1, rBCC2), and (b) Bridges between bit lines (of adjacent cells) belonging to the same or different ports (rBCC3, rBCC4, rBCC5, rBCC6, rBCC7, rBCC8); see Appendix B.

The 3PF2$_v$: This type can be caused by bridges between nodes of adjacent cells belonging to the same row (rBCC1, rBCC2), to the same column (cBCC1, cBCC2), or on the same diagonal (dBCC1, dBCC2), see Appendix B; it consists of two FFMs:

1. wCFrd&wRDF&wRDF: Applying three simultaneous read operations to the v-cell will cause the v-cell to flip if the a-cell is in a certain state. The read operations then return *wrong* values. This FFM consists of four FPs, as shown in Table 8.4.

2. wCFdrd&wDRDF&wDRDF: Applying three simultaneous read operations to the v-cell will cause the cell to flip if the a-cell is in a certain state. The read operations then return *correct* values. This FFM consists of four FPs, as shown in Table 8.4.

It should be noted that the 3PFs discussed above are valid for 3P memories which allow for two simultaneous reads and a write of the same location (i.e., 'wx_c:ry_c:ry_c'). If this is not supported, then the FFM: wCFds&wCFds&wCFds will consist only of the FPs sensitized by three simultaneous read operations to the same location; i.e., '< $xrx : xrx : xrx; 0/1/-$ >$_{a,v}$' and '< $xrx : xrx : xrx; 1/0/-$ >$_{a,v}$' ($x \in \{0,1\}$).

8.4   Fault probabilities analysis

In order to determine the importance of each FFM for 3P memories, their probability of occurrence has to be calculated. This will be done based on the simulation results (i.e., the fault behavior and the corresponding resistance range) under the assumption that all SDs (i.e., opens, shorts, and bridges) are equal likely to occur. In addition, a distribution of the resistance values of defects based on industrial data will be used.

First, the probability of occurrence of each FP (sensitized by an open, short or a bridge) is determined. Since in our simulation, each simulated SD is a representative of a group, the number of defects within a group has to be taken into account. Thereafter, the probabilities of the FPs constituting a certain FFM (e.g., SF) are summed in order to calculate the probability of occurrence of this FFM; remember that a FFM is a non empty set of FPs. We will focus on the relative probability of occurrence of each FFM (i.e., 1PF1 like SF, 2PF1 like wRDF&wRDF and 3PF1 like wRDF&wRDF&wRDF) compared with the other FFMs within all 3P memory cell array FFMs caused by all SDs (i.e., the SFDs and MFDs).

Table 8.5 lists the FFMs that can be sensitized, together with their probabilities. The table shows that the occurrence probability of 1PFs is 80.333%, that of 2PFs is 15.791% and that of 3PFs is 3.876%. That means that if only conventional SP tests are used to test 3P memories, the fault coverage can not be more than 80.333%, which is not acceptable. Therefore, the 2PFs as well as the 3PFs have to be considered. That requires, in addition to SP tests, 2P tests as well as 3P tests.

Table 8.5. Probabilities of the FFMs in 3P memories

Class	Sub. Class	Fault model	Prob. %
1PF (80.333%)	1PF1 (56.759%)	SF	29.315
		TF	10.312
		RDF	4.455
		DRDF	0.448
		IRF	7.231
		RRF	2.507
		UWF	0.978
		URF	0.030
		DRF	0.575
	1PF2 (23.574%)	CFst	11.355
		CFds	6.466
		CFtr	0.806
		CFrd	0.435
		CFdrd	0.097
		CFir	3.373
		CFrr	1.042
2PF (15.791%)	2PF1	wRDF&wRDF	0.944
		wDRDF&wDRDF	0.926
		wRDF&wTF	0.357
	2PF2$_a$	wCFds&wCFds	2.536
	2PF2$_v$	wCFrd&wRDF	0.593
		wCFdrd&wDRDF	0.127
	2PF2$_{av}$	wCFds&wRDF	3.105
		wCFds&wIRF	5.775
		wCFds&wRRF	1.428
3PF (3.876%)	3PF1	wDRDF&wDRDF&wDRDF	0.814
		wRDF&wRDF&wRDF	0.172
	2PF2$_a$	wCFds&wCFds&wCFds	2.142
	2PF2$_v$	wCFrd&wRDF&wRDF	0.197
		wCFdrd&wDRDF&wDRDF	0.574

Another aspect that shows the importance of taking 2PFs as well as 3PFs into consideration is the impact on the Reject Ratio (RR); the RR is the fraction of the defective parts that pass all tests and hence is sold to customers. This ratio, given in terms of defects per million (DPM), can be estimated based on the JSSC model [51]:

$$RR = \frac{(1-FC)*(1-Y)*e^{(-FC.(N_0-1))}}{Y+(1-FC)*(1-Y)*e^{(-FC.(N_0-1))}}$$

whereby FC is the fault coverage; Y is the yield and N_0 is the average number of faults per defective die.

Assume that the yield by taking only 1PFs into consideration is $Y_{1PFs} = 0.95$, and that the yield by taking both 1PFs and 2PFs into considerations is $Y_{1PFs+2PF1s} = 0.95 - 0.95*15.791\% \simeq 0.800$ (note that 15.791% is the percentage of 2PFs). Additionally, assume that the yield by taking 1PFs, 2PFs as well as 3PFs into consideration is $Y_{All} = 0.95 - 0.95 * 19.667\% = 0.763$ (note that 19.667% is the percentage of 2PFs together with 3PFs). Moreover, assume that the fault coverage, when only 1PFs are taken into consideration, is $FC_{1PFs} = 80.333\%$; this is the maximal fault coverage that can be reached with SP tests since their probability is 80.333%. Furthermore, when conventional SP tests as well as special tests for 2PFs will be used, then in addition to 1PFs, assume that 95% of the 2PFs will be detected. That means that the fault coverage will be 80.333% + 0.95 * 15.791% ≃ 95.306%. Finally, when SP tests are used together with special tests for 2PFs and 3PFs, then assume that 95% of 2PFs and 3PFs will be covered, such that the fault coverage will be 80.333% + 0.95 * 19.667% ≃ 99.017%.

Based on the the above formula and assumptions, the RR when only 1PFs are considered (RR_{1PFs}), when both 1PFs and 2PF1s are considered ($RR_{1PFs+2PFs}$), and when 1PFs, 2PFs as well as 3PFs are considered (RR_{All}) can be calculated. Table 8.6 shows RR_{1PFs}, $RR_{1PFs+2PFs}$ and RR_{All} for different values of N_0[2]; in the table, RF denotes the reduction factor for the case all faults are considered. It can be seen from the table that the RF can be reduced with a factor 6.62 on the average.

Table 8.6. Reject ratio reduction as a function of N_0

N_0	RR_{1PFs}	$RR_{1PFs+2PFs}$	RR_{All}	RF
3	2071.669	1741.441	421.253	4.918
4	928.829	672.148	156.543	5.933
5	416.176	259.260	58.163	7.155
6	186.490	99.976	21.609	8.630

---

[2]these values are based on Intel experimental data

8.5   Realistic fault models for p-port memories

In this section a classification of FFMs for p-port (pP) memories will be established. This will be done based on the results found for 2P and 3P memories, which have been derived based on circuit simulation.

The 3PFs are divided into 3PF1s and 3PF2s. Such faults can be considered as an extension of the 2PFs; compare Figure 5.4 and Table 5.6 with Figure 8.2 and Table 8.4. The 3PF1s, which consist of two FFMs, can be considered as an extension of the 2PF1s. For instance, the 3PF1 wRDF&wRDF&wRDF is an extension of the 2PF1 wRDF&wRDF. On the other hand, the introduced 3PF2s are divided into the fault types $3PF2_a$ and $3PF2_v$, which are extensions of the $2PF2_a$, respectively, the $2PF2_v$. By inspecting Figure 5.4 and Figure 8.2, one can see that there is no 3PF that can be considered as an extension of the 2PF1 wRDF&wTF, neither of the $2PF2_{av}$ (i.e., the 2PF2 is sensitized by applying two simultaneous sensitizing operations to two different cells: a-cell and v-cell). Such faults are caused by bridges between bit lines belonging to two different ports (see Section 5.3.2). It has been shown with Inductive Fault Analysis that a bridge only occurs between physically adjacent lines, and that the occurrence probability of bridges involving *at the most* two nodes is very large (96.6% on the average) as compared with bridges involving more than two nodes. Therefore the assumption will be made that the $2PF2_{av}$ can only be caused by bridges involving at most two bit lines (belonging to different ports) that are physically adjacent to each other. That means that, irrespective of the number of ports the MP memory consists of, the bridges between two bit lines belonging to any two different ports can only cause a wRDF&wTF or a $2PF2_{av}$. Therefore, these are unique 2PFs that can not be extended.

Based on the above discussion, the FFMs for any MP memory can be derived. Such faults can be divided, based on the number of ports required in order to be sensitized, into p-classes: single-port faults (1PFs), two-port faults (2PFs), three-port faults (3PFs), ..., p-port faults ($pPFs$); see Figure 8.3. The 1PFs and the 2PFs have been addressed in Chapter 5; while the 3PFs have been established in Section 8.3. Below the $pPFs$ ($p > 3$) will be discussed.

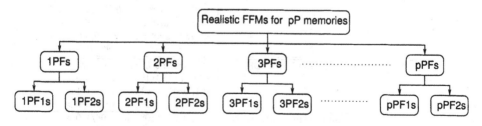

Figure 8.3. Classification of realistic FFMs in pP memories

The pPFs (for $p \geq 3$) are faults that can only be sensitized by applying p simultaneous operations; they are divided into faults involving a single-cell (pPF1s) and faults involving two cells (pPF2s); see Figure 8.4. The pPF1s are based on a combination of p *single-cell weak faults*; while the pPF2s are based on a combination of single-cell weak faults and weak faults involving two cells. The latter are divided into two types (see Figure 8.4): the pPF2$_a$ and the pPF2$_v$. The pPF2$_a$ is sensitized in the v-cell by applying p simultaneous operations to the a-cell c_a (solid arrows in the figure); while the pPF2$_v$ is sensitized by applying p simultaneous operations to the v-cell c_v, while the a-cell c_a has to be in certain state (dashed arrows in the figure).

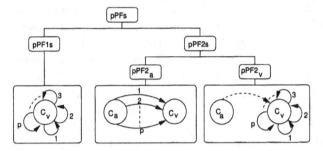

Figure 8.4. Classification of pPFs

A taxonomy of all realistic pPF FFMs is given in Table 8.7, together with their FPs. For the FP notation, a similar explanation can be given as that given for 3PFs. For instance, the '$< 0r0^a : 0r0^b : ... : 0r0^p/1/1 >$' denotes p simultaneous read operations ($0r0^a$ denotes $0r0$ via port P_a) applied to the v-cell cause the v-cell to flip, and the sense amplifiers returns 1.

Table 8.7. List of realistic pPF FFMs; $x \in \{0, 1\}$ and $d =$ don't care

Class	#	FFM	Fault primitives
pPF1	1	wRDF&wRDF...&wRDF	$< 0r0^a : 0r0^b : ... : 0r0^p/1/1 >$, $< 1r1^a : 1r1^b : ... : 1r1^p/0/0 >$
	2	wDRDF&wDRDF...&wDRDF	$< 0r0^a : 0r0^b : ... : 0r0^p/1/0 >$, $< 1r1^a : 1r1^b : ... : 1r1^p/0/1 >$
pPF2$_a$	3	wCFds&wCFds...&wCFds	$< dw0^a : drd^b : ... : drd^p; 0/1/- >$, $< dw0^a : drd^b : ... : drd^p; 1/0/- >$, $< dw1^a : drd^b : ... : drd^p; 0/1/- >$, $< dw1^a : drd^b : ... : drd^p; 1/0/- >$, $< xrx^a : xrx^b : ... : xrx^p; 0/1/- >$, $< xrx^a : xrx^b : ... : xrx^p; 1/0/- >$
pPF2$_v$	4	wCFrd&wRDF...&wRDF	$< 0; 0r0^a : 0r0^b : ... : 0r0^p/1/1 >$, $< 0; 1r1^a : 1r1^b : ... : 1r1^p/0/0 >$, $< 1; 0r0^a : 0r0^b : ... : 0r0^p/1/1 >$, $< 1; 1r1^a : 1r1^b : ... : 1r1^p/0/0 >$
	5	wCFdrd&wDRDF...&wDRDF	$< 0; 0r0^a : 0r0^b : ... : 0r0^p/1/0 >$, $< 0; 1r1^a : 1r1^b : ... : 1r1^p/0/1 >$, $< 1; 0r0^a : 0r0^b : ... : 0r0^p/1/0 >$, $< 1; 1r1^a : 1r1^b : ... : 1r1^p/0/1 >$

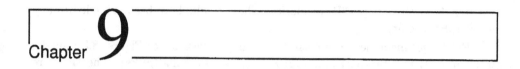

Chapter **9**

Tests for p-port SRAMs

9.1 Condition for detecting p-port faults

9.2 Tests for p-port faults

9.3 Test strategy

Chapter 8 introduced realistic fault models for p-port (pP) memories. They have been divided into p classes: single-port faults (1PFs), two-port faults (2PFs), ..., p-port faults (pPFs). Tests for 1PFs and 2PFs have been addressed in Chapter 6; while tests for pPFs (p > 2) have to be established.

This chapter concerns with developing tests for p-port memories, together with an optimal test strategy. It starts with introducing conditions to detect the pPFs (p > 2) in Section 9.1. They will be used to derive tests in Section 9.2; the tests will be optimized and thereafter merged into a single test, such that all pPFs will be testable with just one single test. Finally, the test strategy, which determines the methodology to be used for testing, will be discussed in Section 9.3.

9.1   Condition for detecting p-port faults

Chapter 5 established realistic functional fault models (FFMs) for 2P memories, based on defect injection and circuit simulation. Similar work has been presented in Chapter 8 for three-port (3P) memories. These results have been extended to any p-port (pP) memory.

FFM for pP memories have been divided into p classes (see Figure 8.3): single-port faults (1PFs), two-port faults (2PFs), ..., p-port faults (pPFs). The 1PFs are faults that occur in conventional SP memories; the 2PFs can not be sensitized using SP operations, they require the use of the two ports of the memory simultaneously; while pPFs are faults requiring the use of the p ports of the memory simultaneously in order to be sensitized. In the remainder of this section, we will focus on developing detection conditions for the pPFs.

The pPFs ($p > 2$) have been divided into pPFs involving a single cell (pPF1s) and pPFs involving two cells (pPF2s); see Figure 8.4. Based on to which the two simultaneous operations have to be applied, in order to sensitize the fault (to the a-cell or to the v-cell), the pPF2s have been further divided into two types: pPF2$_a$ and pPF2$_v$, respectively. The established realistic FFMs for pPFs are regiven in Table 9.1, together with their FPs. In the following, the conditions for detecting such faults will be discussed; first for pPF1s and thereafter for pPF2s. These conditions will be used to derive functional tests. The march notation used to describe the conditions (and tests) has been explained in Section 6.1.

Table 9.1. List of realistic pPF FFMs; $x \in \{0, 1\}$ and $d =$ don't care

Class	#	FFM	Fault primitives
pPF1	1	wRDF&wRDF...&wRDF	$< 0r0^a : 0r0^b : ... : 0r0^p/1/1 >,$ $< 1r1^a : 1r1^b : ... : 1r1^p/0/0 >$
	2	wDRDF&wDRDF...&wDRDF	$< 0r0^a : 0r0^b : ... : 0r0^p/1/0 >,$ $< 1r1^a : 1r1^b : ... : 1r1^p/0/1 >$
pPF2$_a$	3	wCFds&wCFds...&wCFds	$< dw0^a : drd^b : ... : drd^p; 0/1/- >,$ $< dw0^a : drd^b : ... : drd^p; 1/0/- >,$ $< dw1^a : drd^b : ... : drd^p; 0/1/- >,$ $< dw1^a : drd^b : ... : drd^p; 1/0/- >,$ $< xrx^a : xrx^b : ... : xrx^p; 0/1/- >,$ $< xrx^a : xrx^b : ... : xrx^p; 1/0/- >$
pPF2$_v$	4	wCFrd&wRDF...&wRDF	$< 0; 0r0^a : 0r0^b : ... : 0r0^p/1/1 >,$ $< 0; 1r1^a : 1r1^b : ... : 1r1^p/0/0 >,$ $< 1; 0r0^a : 0r0^b : ... : 0r0^p/1/1 >,$ $< 1; 1r1^a : 1r1^b : ... : 1r1^p/0/0 >$
	5	wCFdrd&wDRDF...&wDRDF	$< 0; 0r0^a : 0r0^b : ... : 0r0^p/1/0 >,$ $< 0; 1r1^a : 1r1^b : ... : 1r1^p/0/1 >,$ $< 1; 0r0^a : 0r0^b : ... : 0r0^p/1/0 >,$ $< 1; 1r1^a : 1r1^b : ... : 1r1^p/0/1 >$

9.1.1 Condition for detecting pPF1s

The pPF1 fault subclass has the property that the fault is sensitized in the v-cell by applying p simultaneous operations to the same cell. It has to be considered for pP memories allowing for p simultaneous read operations to the same location. It consists of two FFMs, each with two FPs; see Table 9.1. Such faults are detectable if the following condition is satisfied.

Condition pPF1: Any pPF1 fault will be detected by a march test which contains the two march elements of Case A, and the two march elements of Case B; these four march elements may be combined into three, two, or a single march element.

- Case A
  (to detect $< 0r0^a : 0r0^b : ... : 0r0^p/1/1 >_v$ and $< 0r0^a : 0r0^b : ... : 0r0^p/1/0 >_v$):
  $\updownarrow (..., r0^a : r0^b : ... : r0^p); \updownarrow (r0^a : -^b : ... : -^p, ...)$

- Case B
  (to detect $< 1r1^a : 1r1^b... : 1r1^p/0/0 >_v$ and $< 1r1^a : 1r1^b... : 1r1^p/0/1 >_v$):
  $\updownarrow (..., r1^a : r1^b : ... : r1^p); \updownarrow (r1^a : -^b, ... : -^p)$

The first p simultaneous read operations through the p ports in each march element sensitize and detect the wRDF&wRDF&...&wRDF (e.g., $< 0r0^a : 0r0^b : ... : 0r0^p/1/1 >_v$) and only sensitize the wDRDF&wDRDF&...&wDRDF (e.g., $< 0r0^a : 0r0^b : ... : 0r0^p/1/0 >_v$). The latter will be detected by the second single read operations. Note that the single read operations can be replaced with any number ($\leq p$) of simultaneous read operations (i.e., '$-^i$' which denotes any allowed operation via port P_i, can be replaced with a read operation).

9.1.2 Condition for detecting pPF2s

Depending on to which cells (to the a-cell and/or to the v-cell) the p simultaneous operations are applied, the pPF2 subclass is divided into two types: pPF2$_a$ and pPF2$_v$; see Figure 8.4 and Table 9.1.

Conditions for detecting pPF2$_a$s

The pPF2$_a$ has the property that the fault is sensitized in the v-cell by applying p simultaneous operations to the a-cell. It consists of one FFM (i.e., wCFds&...&wCFds) with eight FPs; see Table 9.1. In order to detect any pPF2$_a$, we have:

1. to apply all sensitizing operations to a the a-cell c_a; $a \in \{0, 1, 2, ..., n-2, n-1\}$.

2. to detect the fault in the v-cell c_v, whereby $v \neq a$, by reading the same cell.

The order in which c_a has to be selected is not relevant. Therefore, the $\Updownarrow_{a=0}^{n-1}$ address order can be specified.

Condition pPF2$_a$: Any pPF2$_a$ is detectable by a march test which contains both march elements of Case A and both march elements of Case B.

- **Case A:**

(to detect $< dw1^a : rd^b : ... : rd^p; 0/1/- >$, $< dw0^a : rd^b : ... : rd^p; 0/1/- >$, and $<rx^a : rx^b : ... : rx^p; 0/1/->$)

- $\Updownarrow_{a=0}^{n-1} (r0^a : r0^b : ... : r0^p, ..., w1^a : rd^b : ... : rd^p, ..., r1^a : r1^b : ... : r1^p, ..., w0^a : rd^b : ... : rd^p)$;
  $\Updownarrow_{a=0}^{n-1} (r0^a : -^b : ... : -^p, ...)$

- **Case B:**

(to detect $< dw1^a : rd^b : ... : rd^p; 1/0/- >$, $< dw0^a : rd^b : ... : rd^p; 1/0/- >$, and $<rx^a : rx^b : ... : rx^p; 1/0/->$)

- $\Updownarrow_{a=0}^{n-1} (r1^a : r1^b : ... : r1^p, ..., w0^a : rd^b : ... : rd^p, ..., r0^a : r0^b : ... : r0^p, ..., w1^a : rd^b : ... : rd^p)$;
  $\Updownarrow_{a=0}^{n-1} (r1^a : -^b : ... : -^p, ...)$

Condition pPF2$_a$ can be explained as follows: the operations in the first march element of Case A will sensitize all pPF2$_a$ faults when the value of the fault effect is 1, because that march element contains all sensitizing operations. If the address order of a is increasing , then Case A will detect the fault by the '$r0^a : r0^b : ... : r0^p$' operation of the first march element when the v-cell has a higher address than the a-cell (i.e., $v > a$), and by '$r0^a : -^b : ... : -^p$' operation of the second march element when $v < a$. If the address order of a is decreasing, then Case A will detect the fault by the '$r0^a : r0^b : ... : r0^p$' operation of the first march element when $v < a$, and by '$r0^a : -^b : ... : -^p$' operation of the second march element $v > a$. Note that the operation '$r0^a : -^b : ... : -^p$' can be replaced with any number of simultaneous read operations fewer than or equal to p. A similar explanation can be given for Case B, which is required to sensitize and to detect pPF2$_a$ faults when the value of the fault effect is 0.

In the above condition, simultaneous $(p-1)$ reads and a write of the same location is assumed to be supported. If this is not the case, then the pPF2$_a$ will consist only of FPs sensitized by p simultaneous read operations; as a consequence the operations '$wx^a : rd^b : ... : rd^p$' in Condition pPF2$_a$ ($x \in \{0, 1\}$) should be replaced with '$wx^a : n^b : ... : n^p$', whereby n denotes no operation.

Conditions for detecting pPF2$_v$s

The pPF2$_v$ has the property that the fault is sensitized in the v-cell by first bringing the a-cell in a certain state, and thereafter applying p simultaneous read operations to the v-cell. It consists of two FFMs, each with four FPs; see Table 9.1. In order to detect the presence of such faults in the v-cell c_v, the following condition has to be satisfied:

Condition pPF2$_v$: Any $pPF2_v$ is detectable by a march test if the test selects all pairs of cells (c_a, c_v) whereby $a \in \{0, 1, ..., v-1, v+1, ..., n-2, n-1\}$; and each pair undergoes the four states 00, 01, 10 and 11. In addition, in each state, p simultaneous read operations, followed by at least a single read operation, have to be applied to the v-cell.

Condition $pPF2_v$ can be explained as follows. When the pair of cells (c_a, c_v) are in the state 00, then the FP '$< 0; 0r0^a : 0r0^b : ... : 0r0^p/1/1 >_{a,v}$' (see Table 9.1) will be sensitized and detected by p simultaneous read operations applied to v-cell; while the FP '$< 0; 0r0^a : 0r0^b : ... : 0r0^p/1/0 >_{a,v}$' will be sensitized. The extra (single) read operation will detect the latter FP. A similar explanation can be given for the other three states, which are required to detect the other six FPs of $pPF2_v$; see Table 9.1.

9.2 Tests for p-port faults

The FFMs for pP memory cell array faults are divided into 1PFs, 2PFs, ..., and pPFs. Therefore the test procedure may be divided into p parts:

1. Test(s) to detect 1PFs.
2. Test(s) to detect 2PFs.
· ...
p. Test(s) to detect pPFs.

Tests for 1PFs are described in literature [2, 21, 24, 88]; while tests for 2PFs have been introduced in Chapter 6. Below, tests for detecting pPFs ($p > 2$) will be developed. First the test for pPF1s will be presented, thereafter for pPF2s. Finally a test detecting all pPFs will be introduced.

9.2.1 Tests for pPF1s

The test shown in Figure 9.1, referred as *March pPF1*, detects all pPF1 faults. March pPF1 consists of four march elements with a total test length of $6n$, whereby n is the size of the memory cell array. Note that the four march elements of the test can be merged into one, two, or three march elements; and that the single read operations (e.g., '$r0^a : -^b : ... : -^{p}$') can be replaced with any number ($\leq p$) of simultaneous read operations.

March pPF1 detects all pPF1 faults since it satisfies Condition pPF1 of Section 9.1.1. The second march element of the test (i.e., M_1) contains the two march

$$\{ \, \updownarrow (w0^a : n^b : ... : n^p) \, ; \quad \updownarrow (r0^a : r0^b : ... : r0^p, r0^a : -^b : ... : -^p) \, ;$$
$$M_0 \qquad\qquad\qquad\qquad M_1$$
$$\updownarrow (w1^a : -^b : ... : -) \, ; \quad \updownarrow (r1^a : r1^b : ... : r1^p, r1^a : -^b : ... : -^p) \, \}$$
$$M_2 \qquad\qquad\qquad\qquad M_3$$

Figure 9.1. March pPF1

elements of case A, while M_3 contains the two march elements of Case B.

9.2.2   Tests for pPF2s

The pPF2s are divided, depending on the cells to which the simultaneous operations are applied, into two types: pPF2$_a$ and pPF2$_v$. Below, tests for each type will be be introduced.

Tests for the pPF2$_a$ faults

March pPF2$_a$, shown in Figure 9.2, detects all pPF2$_a$ faults since it satisfies Condition pPF2$_a$ of Section 9.1.2. M_1 and M_2 contain the two march elements of Case A, while M_3 and M_4 contain the two march elements of Case B. March pPF2$_a$ ($p > 2$) has a test length of $12n$, which is the same as the test length of March 2PF2$_a$ introduced in Chapter 6. Note that March pPF2$_a$ can also detect the pPF1: wRDF&wRDF&...&wRDF.

$$\{ \, \updownarrow (w0^a : n^b : ... : n^p) \, ;$$
$$M_0$$
$$\updownarrow (r0^a : r0^b : ... : r0^p, w1^a : r0^b : ... : r0^p, r1^a : r1^b : ... : r1^p, w0^a : r1^b : ... : r1^p) \, ;$$
$$M_1$$
$$\updownarrow (r0^a : -^b : ... : -, w1^a : -^b : ... : -^p) \, ;$$
$$M_2$$
$$\updownarrow (r1^a : r1^b : ... : r1^p, w0^a : r1^b : ... : r1, r0^a : r0^b : ... : r0^p, w1^a : r0^b : ... : r0^p) \, ;$$
$$M_3$$
$$\updownarrow (r1^a : -^b : ... : -^p) \, \}$$
$$M_4$$

Figure 9.2. March pPF2$_a$

March pPF2$_a$ can be further optimized to a $10n$ test without impacting the fault coverage of the pPF2$_a$ faults. The result is shown in Figure 9.3, and referred as *March pPF2$_a$-*. Note that the addressing sequence of March pPF2$_a$- is relevant; it is important for the fault coverage. Although the test of Figure 9.7 does not contain explicitly the two pairs of march elements of Condition pPF2$_a$, it detects all pPF2$_a$ faults; the sensitizing/detection operations for pPF2$_a$ faults are distributed over M_1, M_2, M_3,

M_4, and M_5. The $pPF2_a$ faults consist of one FFM: wCFds&wCFds&...&wCFds with the eight FPs; see Table 9.1.

- The $< 0r0^a : 0r0^b : ... : 0r0^p; 0/1/- >_{a,v}$ and $< dw1^a : drd^b : ... : drd^p; 0/1/- >_{a,v}$ will be sensitized and detected by M_1 if the v-cell has a higher address than a-cell; i.e., $v > a$. If $v < a$, then these faults will be sensitized and detected by M_3.

- The $< 1r1^a : 1r1^b : ... : 1r1^p; 1/0/- >_{a,v}$ and $< w0^a : drd^b : ... : drd^p; 1/0/- >_{a,v}$ will be sensitized and detected by M_2 if $v > a$. If $v < a$, then these faults will be sensitized and detected by M_4.

- The $< 0r0^a : 0r0^b : ... : 0r0^p; 1/0/- >_{a,v}$ and $< w1^a : xrx^b : ... : xrx^p; 1/0/- >_{a,v}$ will be sensitized by M_1 and detected by M_2 if $v < a$; while the same faults will be sensitized by M_3 and detected by M_4 if $v > a$.

- The $< 1r1^a : 1r1^b : ... : 1r1^p; 0/1/- >_{a,v}$ and $< w0^a : drd^b : ... : drd^p; 0/1/- >_{a,v}$ will be sensitized by M_2 and detected by M_3 if $v < a$; while the same faults will be sensitized by M_4 and detected by M_5 if $v > a$.

$$
\begin{array}{l}
\{ \, \updownarrow (w0^a : n^b : ... : n^p) \; ; \\
\qquad M_0 \\
\Uparrow (r0^a : r0^b : ... : r0^p, w1^a : r0^b : ... : r0^p) \; ; \quad \Uparrow (r1^a : r1^b : ... : r1^p, w0^a : r1^b : ... : r1^p) \; ; \\
\qquad M_1 \qquad\qquad\qquad\qquad\qquad\qquad\qquad\qquad M_2 \\
\Downarrow (r0^a : r0^b : ... : r0^p, w1^a : r0^b : ... : r0^p) \; ; \quad \Downarrow (r1^a : r1^b : ... : r1^p, w0^a : r1^b : ... : r1^p) \; ; \\
\qquad M_3 \qquad\qquad\qquad\qquad\qquad\qquad\qquad\qquad M_4 \\
\Downarrow (r0^a : -^b : ... : -^p) \; \} \\
\qquad M_5
\end{array}
$$

Figure 9.3. March $pPF2_a$-; the optimized version of March $pPF2_a$

It should be noted that for March $pPF2_a$, and its optimal version, it is assumed that $(p - 1)$ simultaneous reads and one write of the same location is allowed in the to be tested pP memory. If this is not the case, then all operations 'wx:ry:...:ry' in Figure 9.2 and Figure 9.3 should be replaced with 'wx : n : ... : n'; whereby $x, y \in \{0, 1\}$.

Tests for the $pPF2_v$ faults

The $pPF2_v$ consists of two FFMs, each with four FPs; see Table 9.1. The test detecting these faults is shown in Figure 9.4, and is referred as *March pPF2_v*; it has a test length of $14n$, which is identical to that of March $2PF2_v$ introduced in Chapter 6. March $pPF2_v$ satisfies Condition $2PF2_v$ of Section 9.1.2. M_0 initializes all memory cells to 0; which means that the state of all pairs (c_a, c_v) is 00. In the second march element, the v-cell is first read via the p ports simultaneously while the state of all pairs (c_a, c_v) is 00; therefore, the FP '$< 0; 0r0^a : 0r0^b : ... : 0r0^p/1/1 >_{a,v}$' (of wCFrd&wRDF...&wRDF) will be detected, while the FP '$< 0; 0r0^a : 0r0^b : ... : 0r0^p/1/0 >_{a,v}$' (of wCFdrd&wDRDF...&wDRDF) will be sensitized. The latter will be detected by the next read operation within the same march element. Thereafter, the v-cell will be written with 1, and then will be read via the p ports simultaneously. This means that p simultaneous read operations are applied to the v-cell, while the state of all pairs (c_a, c_v) is 01; therefore the FP '$< 0; 1r1^a : 1r1^b : ... : 1r1^p/0/0 >_{a,v}$' will be detected, while the FP '$< 0; 1r1^a : 1r1^b : ... : 1r1^p/0/1 >_{a,v}$' will be sensitized. The latter fault will be detected by the next read operation within the same march element. Finally, the v-cell is written with 0, such that all pairs (c_a, c_v) again enter state 00. A similar explanation can be given for march elements M_2 and M_3. Note that March March $pPF2_v$ also detects 2PF1 faults since it satisfies Condition 2PF1 of Section 9.1.1: M_1 (also M_3) contains the march elements of Case A and Case B merged into a single march element.

$$\begin{aligned}
\{ &\updownarrow (w0^a : n^b : ... : n^p) ; \\
&\qquad M_0 \\
\updownarrow &(r0^a : r0^b : ... : r0^p, r0^a : {-}^b : ... : {-}^p, w1^a : {-}^b : ... : {-}^p, \\
&\ r1^a : r1^b : ... : r1^p, r1^a : {-}^b : ... : {-}^p, w0^a : {-}^b : ... : {-}^p); \\
&\qquad M_3; \\
\updownarrow &(w1^a : {-}^b : ... : {-}^p) ; \\
&\qquad M_2 \\
\updownarrow &(r1^a : r1^b : ... : r1^p, r1^a : {-}^b : ... : {-}^p, w0^a : {-}^b : ... : {-}^p, \\
&\ r0^a : r0^b : ... : r0^p, r0^a : {-}^b : ... : {-}^p, w1^a : {-}^b : ... : {-}^p)\} \\
&\qquad M_3
\end{aligned}$$

Figure 9.4. March $pPF2_v$

March $pPF2_v$ can also be optimized without impacting the fault coverage of the $pPF2_v$ faults. The result is shown in Figure 9.5, and referred as *March $pPF2_v$-*; it has a test length of $13n$, which is the same test length as that of March $2PF2_v$-introduced in Chapter 6. Note that the addressing sequence of March $pPF2_v$- is relevant; it is important for the fault coverage.

March $pPF2_v$- detects all $pPF2_v$ faults since Condition $pPF2_v$ of Section 9.1.2 is satisfied. Table 9.2 shows the operations performed on two cells c_i and c_j by march elements of Figure 9.5. The table contains a column 'State' which identifies the state $S_{i,j}$ before the operation, and a column 'State $S_{i,j}$' which identifies the state after the operation. The table shows that all states of (c_i, c_j) (i.e., 00, 01, 11, 10) are generated, and in each state p simultaneous read operations followed by (at least a single) read operation are applied to cells c_i and c_j.

It should be noted that the single read operations (e.g., "$r0^a : {-}^b : ... : {-}^p$") in March $pPF2_v$ and March $pPF2_v$- can be replaced with any number ($\leq p$) of

$$\{ \; \updownarrow (w0^a : n^b : ... : n^p)) \; ;$$
$$M_0$$
$$\Uparrow (r0^a : r0^b : ... : r0^p, r0^a : -^b : ... : -^p, w1^a : -^b : ... : -^p) \; ;$$
$$M_1$$
$$\Uparrow (r1^a : r1^b : ... : r1^p, r1^a : -^b : ... : -^p, w0^a : -^b : ... : -^p) \; ;$$
$$M_2$$
$$\Downarrow (r0^a : r0^b : ... : r0^p, r0^a : -^b : ... : -^p, w1^a : -^b : ... : -^p) \; ;$$
$$M_3$$
$$\Downarrow (r1^a : r1^b : ... : r1^p, r1^a : -^b : ... : -^p, w0^a : -^b : ... : -^p) \; \}$$
$$M_4$$

Figure 9.5. March $pPF2v$-; the optimized version of March $pPF2_v$

simultaneous read operations.

Table 9.2. State table for detecting $pPF2_v$ faults

Step	March element	State	Operation	State $S_{i,j}$
1	M_0	$--$	'$w0^a : n^b : ... : n^p$' to c_i	$0-$
2		$0-$	'$w0^a : n^b : ... : n^p$' to c_j	00
3	M_1	00	'$r0^a : r0^b : ... : r0^p$' to c_i	00
4		00	'$r0^a : -^b : ... : -^p$' to c_i	00
5		00	'$w1^a : -^b : ... : -^p$' to c_i	10
6		10	'$r0^a : r0^b : ... : r0^p$' to c_j	10
7		10	'$r0^a : -^b : ... : -^p$' to c_j	10
8		10	'$w1^a : -^b : ... : -^p$' to c_j	11
9	M_2	11	'$r1^a : r1^b : ... : r1^p$' to c_i	11
10		11	'$r1^a : -^b : ... : -^p$' to c_i	11
11		11	'$w0^a : -^b : ... : -^p$' to c_i	01
12		01	'$r1^a : r1^b : ... : r0^p$' to c_j	01
13		01	'$r1^a : -^b : ... : -^p$' to c_j	01
14		01	'$w0^a : -^b : ... : -^p$' to c_j	00
15	M_3	00	'$r0^a : r0^b : ... : r0^p$' to c_j	00
16		00	'$r0^a : -^b : ... : -^p$' to c_j	00
17		00	'$w1^a : -^b : ... : -^p$' to c_j	01
18		01	'$r0^a : r0^b : ... : r0^p$' to c_i	01
19		01	'$r0^a : -^b : ... : -^p$' to c_i	01
20		01	'$w1^a : -^b : ... : -^p$' to c_i	11
21	M_4	11	'$r1^a : r1^b : ... : r1^p$' to c_j	11
22		11	'$r1^a : -^b : ... : -^p$' to c_j	11
23		11	'$w0^a : -^b : ... : -^p$' to c_j	10
24		10	'$r1^a : r1^b : ... : r1^p$' to c_i	10
25		10	'$r1^a : -^b : ... : -^p$' to c_i	10
26		10	'$w0^a : -^b : ... : -^p$' to c_i	00

9.2.3   Test for all pPFs

By inspecting the three tests, March pPF1, March pPF2$_a$ and March pPF2$_v$, we can see that they are *single-addressing tests*; i.e., they access one cell at a time. This property makes it easy to merge the three tests in a single march test. The result is shown in Figure 9.6 and referred as *March spPF*. The test satisfies Conditions pPF1 of Section 9.1.1 by M_1 as well as by M_3. It also satisfies Condition pPF2$_a$ and Condition pPF2$_v$ of Section 9.1.2. Condition pPF2$_a$ is satisfied by M_1, M_2, M_4, and M_5. M_1 and M_2 contain the two march elements of Case A, while M_4 and M_5 contain the two march elements of Case B. On the other hand, Condition pPF2$_v$ is satisfied by M_1 and M_4. Note that M_1 and M_4 are the same as M_1 and M_3 of March pPF2$_v$ (see Figure 9.4), except that, e.g., '$w0^a : -^b : ... : -^p$' is replaced with '$w0^a : r1^b : ... : r1^p$'; this does not impact the fault coverage.

Figure 9.6 shows that March spPF has a test length of $16n$; while the test lengths of March pPF1, March pPF2$_a$ and March pPF2$_v$ are $6n$, $12n$, and $14n$, respectively. Therefore, in order to detect pPF1, pPF2$_a$ and pPF2$_v$ faults, one can use March spPF instead of testing these faults separately. This will reduce the test time with 50% (i.e., a reduction of the total test length from $6n + 12n + 14n = 32n$ to $16n$).

$$
\begin{array}{l}
\{ \ \updownarrow (w0^a : n^b : ... : n^p) \ ; \\
\qquad M_0 \\
\updownarrow (r0^a : r0^b : ... : r0^p, r0^a : -^b : ... : -^p, w1^a : r0^b : ... : r0^p, \\
\quad r1^a : r1^b : ... : r1^p, r1^a : -^b : ... : -^p, w0^a : r1^b : ... : r1^p); \\
\qquad\qquad M_1 \\
\updownarrow (r0^a : -^b : ... : -^p) \ ; \ \ \updownarrow (w1^a : -^b : ... : -^p) \ ; \\
\qquad M_2 \qquad\qquad\qquad M_3 \\
\updownarrow (r1^a : r1^b : ... : r1^p, r1^a : -^b : ... : -^p, w0^a : r1^b : ... : r1^p, \\
\quad r0^a : r0^b : ... : r0^p, r0^a : -^b : ... : -^p, w1^a : r0^b : ... : r0^p); \\
\qquad\qquad M_4 \\
\updownarrow (r1^a : -^b : ... : -^p) \ \} \\
\qquad M_5
\end{array}
$$

Figure 9.6. March spPF for all pPFs

March spPF can be further optimized by splitting M_1 and M_4 each into two march elements. Figure 9.7 shows the optimized version of March spPF, referred as *March spPF-*. It consists of six march elements and has a test length of $14n$; i.e., $2n$ less then March spPF. That means that if this version is used to test pPF1, pPF2$_a$ and pPF2$_v$ faults, then the test time reduction will be 56.25% (i.e., a test length reduction from $32n$ to $14n$), compared with the case the faults are tested separately. Note that the addressing sequence of the optimized version is *relevant*; it is important for the fault coverage.

March spPF- detects all pPF1s since it satisfies Condition pPF1 of Section 9.1.1:

$$\{ \; \Updownarrow (w0^a : n^b : ... : n^p)) \; ;$$
$$M_0$$
$$\Uparrow (r0^a : r0^b : ... : r0^p, r0^a : -^b : ... : -^p, w1^a : r0^b : ... : r0^p) \; ;$$
$$M_1$$
$$\Uparrow (r1^a : r1^b : ... : r1^p, r1^a : -^b : ... : -^p, w0^a : r1^b : ... : r1^p) \; ;$$
$$M_2$$
$$\Downarrow (r0^a : r0^b : ... : r0^p, r0^a : -^b : ... : -^p, w1^a : r0^b : ... : r0^p) \; ;$$
$$M_3$$
$$\Downarrow (r1^a : r1^b : ... : r1^p, r1^a : -^b : ... : -^p, w0^a : r1^b : ... : r1^p) \; ;$$
$$M_4$$
$$\Downarrow (r0^a : r1^b : ... : r1^p) \; \}$$
$$M_5$$

Figure 9.7. March spPF-; the optimized version of March spPF

M_1 (also M_3) contains the two march element of case A, while M_2 (also M_4) contains the two march element of case B. In addition, it detects all pPF2$_a$ faults; note that the march elements of March spPF2- are the same as those of March pPF2$_a$- (see Figure 9.3); except that the elements M_1, M_2, M_3, and M_4 are extended with read operations (that do not impact the fault coverage). Moreover all pPF2$_v$ faults will be detected since the march elements M1 through M_4 of March spPF- are the same as those of March pPF2$_v$ (see Figure 9.5), with the only difference that '$w0^a : -^b : ... : -^p$' in M_2 and M_4 is replaced with '$w0^a : r1^b : ... : r1^p$'; and '$w1^a : -^b : ... : -^p$' in M_1 and M_3 with '$w1^a : r0^b : ... : r0^p$'; this does not impact the fault coverage.

It should be noted that for March spPF and its optimal version, it is assumed that $(p - 1)$ simultaneous reads and a write of the same location is allowed in the to be tested pP memory. If this is not the case, then all operations '$wx^a : ry^b : ... : ry^p$' in Figure 9.6 and Figure 9.7 should be replaced with '$wx^a : n^b : ... : n^p$'; whereby $x, y \in \{0, 1\}$.

9.2.4   Summary of pP tests

Table 9.3 summarizes the tests introduced in this chapter. It shows their required number of operations (i.e., test length) including the initialization, together with their fault coverage; see also Table 9.1.

It will be clear from the above that all pPFs ($p > 2$) in pP memories can be detected with a *linear* march test; namely March spPF with a test length of $16n$, or with March spPF- with a test length of $14n$. This is very attractive industrially.

9.3   Test strategy

In has been shown in the previous section that all pPFs, for $p \geq 3$, require tests with a time complexity of order $\Theta(n)$. In addition, it has been shown in Chapter 6 that

Table 9.3. Summary of the pP tests; $p > 2$

Test	Test length	Fault coverage
March pPF1	$6n$	All pPF1s
March pPF2$_a$	$12n$	All pPF2$_a$s The pPF1s: wRDF&wRDF&...&wRDF
March pPF2$_a$-	$10n$	All pPF2$_a$s The pPF1s: wRDF&wRDF&...&wRDF
March pPF2$_v$	$14n$	All pPF2$_v$s, all pPF1s
March pPF2$_v$-	$13n$	All pPF2$_v$s, all pPF1s
March spPF	$16n$	All pPF1s, all pPF2$_a$s, and all pPF2$_v$s
March spPF-	$14n$	All pPF1s, all pPF2$_a$s, and all pPF2$_v$s

2PFs require only linear tests. The question that arises now is the following: In order to test a p-port memory, do we need to test each pPF class (i.e., 1PF, 2PF, 3PF, etc) separately? That is apply:

1. Test(s) to detect 1PFs p times.

2. Test(s) to detect 2PFs $C_2^p = \frac{p^2-p}{2}$ times.

3. Test(s) to detect 3PFs C_3^p times.

.. ....

p. Test(s) to detect $pPFs$ once.

The answer to the above question is "*no*". The above test procedure can be optimized by taking into consideration the nature of each pPF class; this will be discussed below.

The pPF class consists of $pPF1s$ and $pPF2s$. The $pPF1s$ for $p > 2$ consist of two FFMs that are extensions of two FFMs of 2PF1s; see Figure 5.4 and Figure 8.4. The sensitization of the $pPF1s$ for $p > 2$ requires the application of p simultaneous read operations to the same location. This will also sensitize 2PF1s, 3PF1s, ... , and $(p-1)$PF1s; except the 2PF1 wRDF&wTF, since that fault is a unique 2PF, which has no extensions for $pPFs$ with $p > 2$. Therefore, a test detecting $pPF1s$ will also detect all $(p-1)$PF1s, ..., 3PF1s, and 2PF1s; except wRDF&wTF. The latter fault, caused by bridges between bit lines belonging to the same column and to *two different ports*, is sensitized by applying a simultaneous read and write to the same location using the two ports; the write operation will fail due to the defect. The first assumption is to apply a test for such faults C_2^p times. However, this can be reduced only to p times as follows:

1. Apply a test detecting wRDF&wTF by performing a write operation via the first port (P_a), and read operations via the other $(p-1)$ ports. In that case, the fault will be detected if it is caused by a bridge between bit lines of P_a and any port $P_i \neq P_a$.

2. Apply a test detecting wRDF&wTF by performing a write operation via P_b, and read operations via the other $(p-1)$ ports. In that case, the fault will be detected if it is caused by a bridge between bit lines of port P_b and any port $P_i \neq P_b$.

...

p. Apply a test detecting wRDF&wTF by performing a write operation via P_p, and read operations via the other $(p-1)$ ports. In that case, the fault will be detected if it is caused by a bridge between bit lines of port P_p and any port $P_i \neq P_p$.

On the other hand, pPF2s for $p > 2$ are divided into pPF2$_a$s and pPF2$_v$s, which are extensions of 2PF2$_a$ and 2PF2$_v$; see Figure 5.4 and Figure 8.4. The sensitization of the pPF$_a$ requires the application of p simultaneous operations to the a-cell. This will also sensitize 2PF2$_a$, 3PF2$_a$, ... , and $(p-1)$PF2$_a$. A similar explanation can be given for pPF2$_v$. Therefore, a test detecting pPF2$_a$ will also detect all $(p-1)$PF2$_a$, ..., 3PF2$_a$, and 2PF2$_a$s; while a test detecting pPF2$_v$ will also detect all $(p-1)$PF2$_v$, ..., 3PF2$_v$, and 2PF2$_v$. Since the 2PF2$_{av}$ faults have no extension for pPFs (see Figure 5.4 and Figure 8.4); they are unique 2PFs and require to be considered separately. Such faults are caused by bridges between bit lines belonging to *two different ports*, to the same (or adjacent) column(s). Their detection requires the application of a write operation to the a-cell and a read operation to the v-cell simultaneously. In order to detect the 2PF2$_{av}$ faults in a p-port memory, the first assumption is to apply a test for such faults C_2^p times. However, this can be reduced to p times; this can be done in a similar way as for wRDF&wTF.

Based on the above, one can conclude that testing a p-port memory can be done by applying:

1. A test(s) to detect 1PFs p times.

2. A test(s) to detect pPFs with $p > 1$ once;
   this includes pPF1s (except wRDF&wTF), pPF2$_a$s and pPF2$_v$s.

3. A test(s) to detect the wRDF&wTF faults p times.

4. A test(s) to detect the 2PF2$_{av}$ faults p times.

It should be clear from the above that the test procedure for a MP memory has a time complexity of $\Theta(p.n)$, whereby p is the number of ports and n is the size of the memory cell array. Since all required tests in the above test procedure are known, an explicit test strategy, together with its test length can be derived; see Figure 9.8. To detect 1PFs, March SS [30] is used. To detect all pPFs, except the wRDF&wTF and the $2PF2_{av}$, March spPF- is used. To detect the $2PF2_{av}$ faults, March d2PF- of Figure 6.12 can be used. To detect wRDF&wTF, a test can be written easily. The following $5n$ test detects all wRDF&wTF faults (see also Condition wRDF&wTF of Section 6.3.1):

$$\{\Updownarrow (w0:n); \Updownarrow (w1:r0,r1:-); \Updownarrow (w0:r1,r0:-)\}$$

The $5n$ test and March d2PF- have to applied in the way explained above. The test strategy of Figure 9.8 will then require a total test length of $(14+37p)n$ since:

- March SS with a test length of $22n$ is applied via each port; i.e., p times.
- March spPF- with a test length of $14n$ is applied once (via p ports).
- March d2PF- with a test length of $10n$ is applied p times.
- The $5n$ test detecting wRDF&wTF faults is applied p times.

Therefore, the total test length for 3P memories, using the strategy of Figure 9.8, will be $89n$.

```
Detection of 1PFs:
    Apply single port test(s) (e.g., March SS) via Pᵢ, i ∈ {a, b, ...,p};
    If die fails, go to FAIL;
Detection of pPFs:
    Apply March spPF- via the p ports;
    If die fails, go to FAIL;
Detection of wRDF&wTF:
    Apply the 5n test p times;
    If die fails, go to FAIL;
Detection of 2PF2ₐᵥs:
    Apply March d2PF- p times;
    If die fails, go to FAIL;
Pass: Print 'Die passes';
      END;
Fail: Print 'Die fails';
      END;
```

Figure 9.8. Test strategy for pP memories

## Chapter 10

Testing restricted p-port SRAMs

10.1 Classification of p-port memories

10.2 Realistic faults for restricted p-port memories

10.3 Tests for restricted p-port memories

10.4 Test strategy for restricted p-port memories

In Chapter 8 and Chapter 9, realistic fault models together with test algorithms for p-port (pP) memories ($p > 2$) have been introduced. However, these only apply to pP memories having p ports with the read as well as the write capability. The ports the memory consists of can have port restrictions; i.e., the port allows only for read or write operations. These restrictions impact the possible fault models, and hence also the tests, as well as the test strategy

In this chapter, first a classification of pP memories will be given, based on the type of the port they consist of; read-only, write only, or read-write port. Then, the impact of port restrictions on the fault modeling will be stated, and realistic fault models for each pP memory class will be presented. Thereafter, tests detecting such faults will be established. Finally, the test strategy for each class will be covered.

10.1 Classification of p-port memories

As mentioned in Chapter 7, multi-port (MP) memories come in different forms, depending on the type of ports they consist of. Each of the ports may have the capability to be a read-only port (Pro), a write-only port (Pwo), or a read-write port (Prw). The total number of ports $p = \#Prw + \#Pwo + \#Pro$. Therefore, for a pP memory, different types can be distinguished based on the port mix. Table 10.1 lists all possible types; for example, in the second row of the table, if the number of Prw is $p-1$ (i.e., $\#Prw = p-1$), then there are two possibilities: a) $\#Pro=1$ and $\#Pwo=0$, or b) $\#Pro=0$ and $\#Pwo=1$. The total number of pP memory types is (see column 5): $1 + 2 + 3 + ... + p + (p+1)$. Since each type has to have a least one port having a read capability, and one port having a write capability, a pP memory with $\#Pro=p$ or $\#Pwo=p$ does not exit; therefore the total number of pP memory types is:

$$[1 + 2 + 3 + ... + p + (p+1)] - 2 = \frac{(p+1)(p+2)}{2} - 2$$

Table 10.1. The different types of pP memories

Label	# Prw	# Pro	# Pwo	# of MP types
1	p	0	0	1
2	$p-1$	1	0	2
		0	1	
3	$p-2$	2	0	3
		1	1	
		0	2	
..	..	..	..	..
..	..	..	..	..
p+1	0	p	0	$p+1$
		$p-1$	1	
		$p-2$	2	
		..	..	
		1	$p-1$	
		0	p	

Depending on the number of simultaneous read operations they allow for, the pP memory types can be further classified into p classes:

1. *p-read pP memories*: these are memories that allow for p simultaneous read operations; i.e., each of the ports is Prw or Pro.

2. *$(p-1)$-read pP memories*: these are memories that allow at the most for $(p-1)$ simultaneous read operations; i.e., each of the ports is Prw or Pro, except one port which is Pwo.

.. ...

$p-1$. *Two-read pP memories*: these are memories that allow at the most for two simultaneous read operations; i.e., $(p-2)$ of the ports are Pwo, and two ports are Prw and/or Pro.

p. *Single-read pP memories*: these are memories that allow at the most for one read operation; i.e., only one of the p ports has the read capability.

10.2 Realistic faults for restricted p-port memories

The FFMs discussed in Chapter 8 are valid for pP memories with #Prw=p. In this section the impact of port restrictions on the FFMs for pP memories will be derived.

FFMs for all pP memory classes can, similar to those of pP memories with #Prw=p, be divided into 1PFs, 2PFs, 3PFs, ..., and pPFs. The pPFs are port mix dependent since they require p simultaneous operations in order to be sensitized; for instance a fault sensitized by p simultaneous read operations can not be realistic for pP memories with #Prw= $p-1$ and #Pwo=1. The consequences of port restrictions on the 2PFs are discussed in detail in Chapter 7. In the rest of this section, the consequences of port restrictions for the pPFs ($p > 2$) of Table 8.7 for the pP memory classes will be analyzed.

10.2.1 The pPFs for p-read pP memories

Depending on their design, p-read pP memories may, or may not, support $(p-1)$ simultaneous reads and a write of the *same* location (e.g., $w0^a : r1^b : ... : r1^p$); however, in the case that this is allowed, the read data will be discarded (write operation has a high priority).

Table 10.2. The pPF FFMs for p-read pP memories; $x \in \{0,1\}$ and $d =$ don't care

Subclass	#	FFM	Fault primitives
pPF1	1	wRDF&wRDF...&wRDF	$< 0r0^a : 0r0^b : ... : 0r0^p/1/1 >$, $< 1r1^a : 1r1^b : ... : 1r1^p/0/0 >$
	2	wDRDF&wDRDF...&wDRDF	$< 0r0^a : 0r0^b : ... : 0r0^p/1/0 >$, $< 1r1^a : 1r1^b : ... : 1r1^p/0/1 >$
pPF2$_a$	3	wCFds&wCFds...&wCFds	$< dw0^a : drd^b : ... : drd^p; 0/1/- >$, $< dw0^a : drd^b : ... : drd^p; 1/0/- >$, $< dw1^a : drd^b : ... : drd^p; 0/1/- >$, $< dw1^a : drd^b : ... : drd^p; 1/0/- >$, $< xrx^a : xrx^b : ... : xrx^p; 0/1/- >$, $< xrx^a : xrx^b : ... : xrx^p; 1/0/- >$
pPF2$_v$	4	wCFrd&wRDF...&wRDF	$< 0; 0r0^a : 0r0^b : ... : 0r0^p/1/1 >$, $< 0; 1r1^a : 1r1^b : ... : 1r1^p/0/0 >$, $< 1; 0r0^a : 0r0^b : ... : 0r0^p/1/1 >$, $< 1; 1r1^a : 1r1^b : ... : 1r1^p/0/0 >$
	5	wCFdrd&wDRDF...&wDRDF	$< 0; 0r0^a : 0r0^b : ... : 0r0^p/1/0 >$, $< 0; 1r1^a : 1r1^b : ... : 1r1^p/0/1 >$, $< 1; 0r0^a : 0r0^b : ... : 0r0^p/1/0 >$, $< 1; 1r1^a : 1r1^b : ... : 1r1^p/0/1 >$

Table 10.2 shows the realistic FFMs for p-read pP memories supporting simultaneous read and write of the same location. They are the same as those shown in Table 8.7.

The pPF1s consist of two FFMs, both require p simultaneous read operations of the same location. Since they allow for such operations, all pPF1s are realistic for p-read pP memories.

The pPF2s consist of two types: pPF2$_a$, and pPF2$_v$. The pPF2$_a$s consist of one FFM, requiring p simultaneous operations to the same location (a-cell); these operations can be p read operations or one write and $(p-1)$ read operations (if the memory allows for such operations). The pPF2$_v$s consist of two FFMs which require p simultaneous read operations to the v-cell. Therefore all pPFs are realistic for the p-read pP memories.

In the case the design of the p-read pP memory does not support simultaneous $(p-1)$ reads and a write to the same location, the pPF2$_a$ FFM (i.e., wCFds&...&wCFds) will consist only of the FPs sensitized by p simultaneous read operations to the same location; that is '$< xrx : xrx : ... : xrx; 0/1/- >_{a,v}$' and '$< xrx : xrx : ... : xrx; 1/0/- >_{a,v}$' ($x \in \{0,1\}$).

10.2.2 The pPFs for $(p-1)$-read pP memories

The $(p-1)$-read pP memories consist of pP memory types that do not support p simultaneous read operations. Therefore, all FFMs sensitized with these operations have to be considered not realistic. In addition, and depending on their design, such memories can allow/not allow for simultaneous $(p-1)$ reads and a write of the *same* location (i.e., '$wx_c^a{:}ry_c^b : ... : ry^p$', $x, y \in \{0, 1\}$).

Table 10.3 show the FFMs for $(p-1)$-read pP memories supporting '$wx_c^a{:}ry_c^b : ... : ry^p$'; see also Table 10.2. Since p simultaneous read operations are not supported, the pPF1s and pPF2$_v$s are not realistic for such pP memories. Only the pPF2$_a$s will be realistic; however, these fault will consist only of FPs sensitized by a write and $(p-1)$ read operations applied to the a-cell simultaneously; none of the FPs are based on p simultaneous read operations to the a-cell. Note that in the case '$wx_c^a{:}ry_c^b : ... : ry^p$' is not supported, the $(p-1)$-read pP memories will have no realistic pPFs.

Table 10.3. The pPF FFMs for $(p-1)$-read pP memories supporting simultaneous write and (p-1) reads to the same location

Subclass	FFM	Fault primitives
pPF1	No fault	-
pPF2$_a$	wCFds&wCFds...&wCFds	$< dw0^a : drd^b : ... : drd^p; 0/1/- >$,
		$< dw0^a : drd^b : ... : drd^p; 1/0/- >$,
		$< dw1^a : drd^b : ... : drd^p; 0/1/- >$,
		$< dw1^a : drd^b : ... : drd^p; 1/0/- >$
pPF2$_v$	No fault	-

10.2.3   The pPFs for other pP memories classes

The other pP memory classes (i.e., single-read pP, two-read pP, ..., $(p-2)$-read pP memories) consist of pP memory types that do not support p simultaneous read operations to the same location. In addition, they don't support a write and $(p-1)$ read operations simultaneously to the same location. Therefore, neither pPF1s, pPF2$_a$s, nor pPF2$_v$s are realistic. The pPFs for such memories is an empty set.

10.3   Tests for restricted p-port memories

The pP tests introduced in Chapter 9 are valid for pP memories with $\#Prw = p$. It has been shown in the previous section that the port restrictions impact the possible set of FFM for each of class of the pP memories; and hence, they also impact the tests. In addition, it has been shown that the pPFs for all classes is an empty set, except for p-read and $(p-1)$-read pP memories. Tests for latter two classes are discussed in the following subsections.

10.3.1   Tests for p-read pP memories

It has been shown in Section 10.2.1 that pPFs for p-read pP memories consist of the same faults as those introduced in Chapter 8, for which a set of tests has been developed in Chapter 9 and summarized in Table 9.3. Therefore, the same tests are applicable to this class of pP memories; March pPF1 to detect the pPF1s; March pPF2$_a$ to detect the pPF2$_a$s, March pPF2$_v$ to detect the pPF2$_v$s, and March sPF (and March sPF-) to detect all the pPFs.

10.3.2   Tests for $(p-1)$-read pP memories

It has been shown in Section 10.2.1 that $(p-1)$-read pP memories have no realistic pPFs if they do not support simultaneous write and $(p-1)$ reads to the same location. As a consequence, no extra test for pPFs is required. However, if this is supported, then the pPFs consists only of pPF2$_a$s with one FFM (see Table 10.3); this FFM consists only of FPs based on simultaneous write and $(p-1)$ read operations to the same location. Therefore, a test to detect such faults is required.

To detect such faults, modified versions of March pPF2$_a$ and March pPF2$_a$- introduced in Section 9.2.2 can be used. The modified version of March pPF2$_a$- is given in Figure 10.1; it has a test length of $10n$. Note that the p simultaneous read operations in the original version are replaced with $(p-1)$ simultaneous reads, which is the maximum number of simultaneous operations allowed in $(p-1)$-read pP memories. These $(p-1)$ simultaneous reads can be replaced with any number of reads (# Reads) whereby $1 \leq \#Reads \leq p-1$; this is because these read operations have the purpose of detecting the faults sensitized by simultaneous write and $(p-1)$ reads operations. In the test given in Figure 10.1, it is assumed that port P_a is the port that does not support the read operations (i.e., Pwo), while all other ports have the read capability (i.e, Prw or Pro).

$$\begin{array}{l}
\{ \ \Updownarrow \ (w0^a : n^b : ... : n^p) \ ; \\
\qquad M_0 \\
\Uparrow (n^a : r0^b : ... : r0^p, w1^a : r0^b : ... : r0^p) \ ; \quad \Uparrow (n^a : r1^b : ... : r1^p, w0^a : r1^b : ... : r1^p) \ ; \\
\qquad\qquad M_1 \qquad\qquad\qquad\qquad\qquad\qquad\qquad M_2 \\
\Downarrow (n^a : r0^b : ... : r0^p, w1^a : r0^b : ... : r0^p) \ ; \quad \Downarrow (n^a : r1^b : ... : r1^p, w0^a : r1^b : ... : r1^p) \ ; \\
\qquad\qquad M_3 \qquad\qquad\qquad\qquad\qquad\qquad\qquad M_4 \\
\Downarrow (n^a : r0^b : ... : r0^p) \ \} \\
\qquad M_5
\end{array}$$

Figure 10.1. March $pPF2_a$- for $(p-1)$-read pP memories

10.4   Test strategy for restricted p-port memories

The test strategy is also impact by the port restrictions. This section describes the impact of the port restrictions on the test strategy for the pP memory classes $(p > 3)$.

10.4.1   Test strategy for p-read pP memories

The p-read pP memories consists of memories having no Pwo ports; all ports are Prw and/or Pro. Assume that $\#Prw = b$ and $\#Pro = r$, whereby $b + r = p$. The test strategy for such memories is shown in Figure 10.2, and is similar to that given for pP memories with $\#Prw = p$ of Figure 9.8.

In order to detect the 1PF1s, a single-port test (e.g., March SS) has to be applied b times; i.e, once via each Prw. Since no test can be applied via Pro, the only possibility is to apply the test in such way that the write operations will be done via any Prw, while the read operations have to be done via each Pro. Note that the total test length to detect 1PFs is $22n \cdot (b + r)$, since the test length of March SS is $22n$.

To detect $pPFs$, March $spPF$- of Figure 9.7 (with a test length of $14n$) has to be applied once via the p ports. The write operations can be applied via any Prw of the memory.

To detect the wRDF&wTF faults, the $5n$ test introduced in Section 9.3 can be used. The test has to be applied by performing a write operation via a Prw and read operations via the other $(p-1)$ ports simultaneously. Since $\#Prw = b$, the test has to be applied b times. This step in the test strategy will require a total test length of $5n.b$.

For the detection of the $2PF2_{av}s$, March $d2PF$- of Figure 6.12 with a test length of $9n$, can be used. The test has to be applied in a similar way as the test for wRDF&wTF faults; i.e., apply March $d2PF$- by performing a write operation via a

Prw and read operations via the other $(p-1)$ ports simultaneously. This test will be also applied b times since $\#Prw = b$. The detection of 2PF2$_{av}$s will require a total test length of $9n.b$.

It will be clear from the above that the test strategy of Figure 10.2 requires a total test length of $14n + n \cdot (36b + 22r)$; whereby $b + r = p$.

```
Detection of 1PFs:
    Apply single port test(s) (e.g., March SS) via each Prw;
    If die fails, go to FAIL;
    Apply single port test(s) (e.g., March SS) via each Pro in such way that:
        Write operations will be performed once via any Prw;
        and read operations will be performed via the Pro;
    If die fails, go to FAIL;
Detection of pPFs:
    Apply March spPF- of Figure  9.7 once via the p ports;
    If die fails, go to FAIL;
Detection of wRDF&wTF:
    Apply the 5n test b times;
    If die fails, go to FAIL;
Detection of 2PF2av s:
    Apply March d2PF- of Figure  6.12 b times;
    If die fails, go to FAIL;
Pass: Print 'Die passes';
    END;
Fail: Print 'Die fails';
    END;
```

Figure 10.2. Test strategy for p-read pP memories

In the case the p-read pP memory does not support simultaneous write and read(s) to the same location, the detection of the of wRDF&wTF will be not required; as a consequence, the total test length will be reduced to $14n + n \cdot (31b + 22r)$.

10.4.2   Test strategy for $(p-1)$-read pP memories

The $(p-1)$-read pP memories are memories for which all ports are Prw and/or Pro; except one port which is Pwo. Assume that $\#Prw = b$, $\#Pro = r$, while $\#Pwo = 1$; note that $b + r + 1 = p$. The test strategy of such memories is shown in Figure 10.3 and is explained below.

In order to detect the 1PFs, March SS can be applied b times; i.e, once via each Prw. Since no test can be applied via Pwo (neither via Pro), the only possibility is to apply the test in such way that the write operations will be done via the Pwo (any Prw) while the read operations have to be done via any Prw (the Pro). Note that the total test length to detect 1PFs is $22n \cdot (b + r + 1)$, since the test length of

```
Detection of 1PFs:
    Apply single port test(s) (e.g., March SS) via each Prw;
    If die fails, go to FAIL;
    Apply single port test(s) (e.g., March SS) via each Pro in such way that:
        Write operations will be performed once via any Prw;
        and read operations will be performed via Pro;
    If die fails, go to FAIL;
    Apply single port test(s) (e.g., March SS) via Pwo in such way that:
        Write operations will be performed via Pwo;
        and read operations will be performed via any Prw or Pro;
    If die fails, go to FAIL;
Detection of pPFs:
    Apply March pPF2_a- of Figure 10.1 once via the p ports;
    If die fails, go to FAIL;
Detection of iPFs:
    Apply March siPF- derived from Figure  9.7 once via the i = b + r = p − 1 ports;
    If die fails, go to FAIL;
Detection of wRDF&wTF:
    Apply the 5n test b+1 times;
    If die fails, go to FAIL;
Detection of 2PF2_av s:
    Apply March d2PF- of Figure  6.12 b+1 times;
    If die fails, go to FAIL;
Pass: Print 'Die passes';
      END;
Fail: Print 'Die fails';
      END;
```

Figure 10.3. Test strategy for $(p-1)$-read pP memories

March SS is $22n$.

The pPFs for $(p-1)$read pP memories $(p > 2)$ consist only of pPF2$_a$s (see Table 10.3), for which the test shown in Figure 10.1 is written. Therefore, in order to detect the pPFs, that test with a test length of $10n$ can be used once via all the ports. The write operations will be applied via Pwo.

On the other hand, since the $(p-1)$read pP memories support $i = b + r = p − 1$ simultaneous reads, the fault class iPF has to be taken into consideration. In that case March siPF- can be derived easily based on March spPF- of Figure 9.7; by replacing the p simultaneous operations with i simultaneous operations. The test length of the test remains the same (i.e., $9n$).

To detect the wRDF&wTF faults, the $5n$ test introduced in Section 9.3 can be used. The test has to be applied by performing a write operation via a Prw or a Pwo, and read operations via the other $(p-1)$ ports simultaneously. Since $b + 1$

ports have the write capability, the test has to be applied $b+1$ times. This step in the test strategy will require a total test length of $5n \cdot (b+1)$.

For the detection of the 2PF2$_{av}$s, March d2PF- of Figure 6.12 can be used. The test has to be applied in a similar way as the test for wRDF&wTF faults; i.e., apply March d2PF- by performing a write operation via a Prw and read operations via the other $(p-1)$ ports simultaneously. This test will be also applied $b+1$ times since $\#Prw = b$ and $\#Pwo = 1$. The detection of 2PF2$_{av}$s will require a total test length of $10n \cdot (b+1)$.

It should be clear form the above that the test strategy of Figure 10.3 requires a total test length of $24n + n \cdot (36b + 22r + 36)$.

10.4.3 Test strategy for other pP memories

The other pP memories, which are i-read pP memories whereby $i < p-1$, consist of memories having two or more Pwo ports. For generality, assume that $\#Prw = b$, $\#Pro = r$, and $\#Pwo = w$. The test strategy of such memories is shown in Figure 10.4 and is explained below.

To detect the 1PFs, the test has to be applied in a similar way as for p-read and $(p-1)$-read pP memories, depending on the type of the ports. Note that the detection of 1PFs requires a total test length of $22n \cdot (b+r+w)$ when using March SS.

In addition, it has been shown in Section 10.2 that the pPFs for memories other then p-read and $(p-1)$-read pP memories $(p > 2)$, consist of an empty set. However, if we assume that the memory support $i = b+r$ simultaneous reads, then the iPFs have to be taken into consideration. In that case March siPF- can be derived easily, based on March spPF- of Figure 9.7, by replacing the p simultaneous operations with i simultaneous operations. The test length of the test remains the same.

Moreover, the unique 2PFs have to be taken into consideration. For such faults (i.e., wRDF&wTF and 2PF2$_{av}$), the $5n$ test of Section 9.3 and March d2PF- of Figure 6.12 have to be used. The tests should be applied $(b+w)$ times, in a similar way as for p-read pP memories.

From the above, one can derive that the total test length of the test strategy of Figure 10.4 is $14n + n \cdot (36b + 36w + 22r)$.

Detection of 1PFs:

   Apply single port test(s) (e.g., March SS) via each Prw;

   If die fails, go to FAIL;

   Apply single port test(s) (e.g., March SS) via each Pro in such way that:

      Write operations will be performed once via any Prw,

      and read operations will be performed via all Pro sequentially;

   If die fails, go to FAIL;

   Apply single port test(s) (e.g., March SS) via each Pwo in such way that:

      Write operations will be performed via Pwo;

      and read operations will be performed via any Prw or Pro;

   If die fails, go to FAIL;

Detection of iPFs:

   Apply March siPF- derived from Figure 9.7 once via the $i = b + w$ ports;

   If die fails, go to FAIL;

Detection of wRDF&wTF:

   Apply the 5n test $b+w$ times;

   If die fails, go to FAIL;

Detection of $2PF2_{av}$s:

   Apply March d2PF- of Figure 6.12 $b+w$ times;

   If die fails, go to FAIL;

Pass: Print 'Die passes';

    END;

Fail: Print 'Die fails';

    END;

Figure 10.4. Test strategy for other pP memories

## Chapter 11

Trends in embedded memory testing

In recent years, embedded memories have become the fastest growing segment of system on chips (SoCs) development infrastructure. According to the International Technology Roadmap for Semiconductors (ITRS 2001), embedded memories will continue to dominate the increasing SoC content in the next years, approaching 94% in about 10 years. Further, the shrinking technology makes memories more sensitive to defects since they are among the most density package modules. Therefore the memory will have a dramatically impact on the overall defect-per-million (DPM) level, hence on the overall SoC yield; i.e., the memory yield will dominate the SoC yield. Meeting a high memory yield requires understanding memory designs, modeling their faulty behaviors in the presence of defects, designing adequate tests and diagnosis strategies as well as efficient repair schemes. This chapter presents the state of art in memory testing including fault modeling, test design, Built-In-Self-Test (BIST) and Built-In-Self-Repair (BISR). Further research challenges and opportunities are discussed in enabling testing (embedded) memories, which use deep submicron technologies.

11.1   Introduction

The advancement in semiconductor IC and technology scaling enhances the integration of more than one function into a single chip of hundreds of transistors and expected to move quickly to billions transistors. The integration of complex systems into a single chip, which may include parts like logic, SRAMs, DRAMs, Flash and analog, is becoming a reality; hence creating system on chips (SoCs).

According to the 2001 ITRS, today's system on chips (SoCs) are moving from logic dominant chips to memory dominant devices in order to deal with today's and future application requirements. Figure 11.1 shows how the dominating logic (about 64% in 1999) is changing in dominating memory (more than 52% today). In addition, SoCs are expected to embed memories of increasing sizes, e.g. 256Mbits and more. As a result, the overall SoC yield will be dominated by the memory yield. Due to the fact that memory yield decreases with the increasing amount of memory, the overall yield may become unacceptable, unless special measures have been taken. The bottom curve in Figure 11.2 shows how the increase in the memory sizes can impact the yield. For instance, the yield of 20MB of embedded memory is about 35%; the example assumes a chips size of 12mm x 12mm, with a memory defect density of 0.4/square-inch and logic defect density of 0.4/square-inch, in 0.13 micron technology. To ensure/enhance an optimal yield level (upper curve in Figure 11.2), embedded memories must have the repair capabilities; hence "repair" is a must for today and future memory technologies. The most know way to deal with this problem is using redundancy or spare elements during the manufacturing [20]; the spare elements can be programmed to replace the faulty cells in the memory matrix. Detecting the faulty dies only is no longer sufficient for SoCs; diagnosis and repair algorithms are often required. The latter form a challenge since it has been shown to be an NP hard problem [45]. A repair algorithm uses a binary fault bit-map as its input. Such bit-map has to be produced based on the used test/diagnosis patterns to catch/locate defect cells. For embedded memories, test pattern(s) is generally programmed inside the BIST engine due to the lack of the controllability of their inputs and the observability of their outputs. The memory test patterns have to guarantee very high defect coverage, otherwise defective memories may escape; hence increasing the DPM level. The quality of the test patterns in terms of the defect coverage (and test length) strongly depends on the used fault models. New memory scaling technologies and processes are introducing new defects that were unknown in the past, and therefore new fault models are emerging.

This all clarifies that the challenges in embedded SoC memory testing will be driven by the following items:

- Fault modeling: New fault models should be established in order to deal with the new defects introduced by today and future (deep-submicron) technologies.

- Test algorithm design: Optimal test/diagnosis algorithms to guarantee high

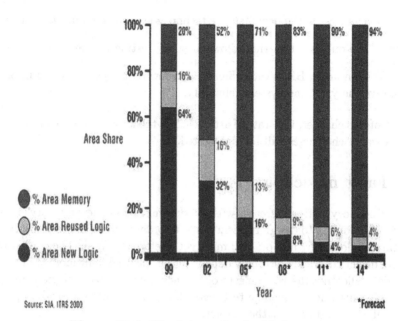

Figure 11.1. The future of embedded memory

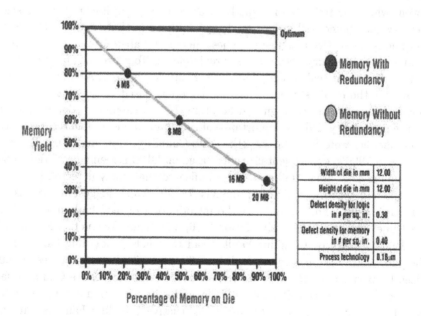

Figure 11.2. Memory sizes versus yield

defect coverage for the new memory technologies and to reduce the DPM level.

- BIST: The only solution that allows at-speed testing for embedded memories.

- BISR: Combining BIST with efficient and low cost repair schemes in order to improve the yield and system reliability.

In the rest of this chapter, the state of art of each of the above issues will be discussed, and the research challenges will be highlighted.

11.2 Fault modeling

The cost of memory testing increases with every generation of new memory chips [35]. Precise fault modeling to design efficient tests, in order to keep the test cost and test time within economically acceptable limits, is therefore essential. The quality of test in terms of defect coverage is strongly dependent on the used fault models. Therefore, fault models reflecting the real defects of the new memory technologies are a must for developing high defect overage test algorithms and therefore providing products with low DPM level driven by the market.

During early 1980's many functional fault models for memories have been introduced. The advantage of such model is that the fault coverage of a certain test can be proven, while the test time is usually of order $O(n)$; i.e., linear with the size of the memory cell. Some well known models introduced in that time are [88]: Stuck-at-Fault and Address-Decoder-Faults. These models are abstract models and are not based on any real memory design and/or real defects. To reflect the faulty behavior of the real defects in real design, Inductive Fault Analysis (IFA) was introduced. IFA allows for the establishment of the fault models based on simulated defects in real physical layout of the design. In addition, IFA is capable of determining the occurrence probability and the importance of each fault models. The result was that new fault models were introduced [21]: State-Coupling Fault and Data-Retention Fault. Late in 1990's, experimental results based on DPM screening of large number of tests applied to a large number of chips indicated that many detected faults cannot be explained with the well known models [75, 90], which suggested the existence of additional faults. This stimulated the introduction of new fault models, based on defect injection and SPICE simulation [5, 6, 24]: Read Destructive Fault, Write Disturb Fault, Transition Coupling Fault, Read Destructive Coupling Fault, etc.

The published work on memory fault modeling described above focuses on faults sensitized by performing *at most one operation*. For instance, State Coupling fault is sensitized in a victim cell if and only if the aggressor cell is in a certain state (i.e., number of required operation is 0); Read Destructive Coupling Fault is sensitized by applying a read operation to the victim cell while the aggressor cell is put in a certain state (i.e., the required number of operation is 1). Memory faults sensitized by performing at the most one operation are referred as *static faults*.

11.2.1 Dynamic fault models

Recent published work reveals the existence of another class of faults in the new memory technologies. It was shown that another kind of faulty behavior can take place in the absence of static faults [7, 33, 32]. This faulty behavior, called *dynamic fault*, requires *more than one operation* to be performed *sequentially* in time in order to be sensitized. For example, a write 1 operation followed immediately by a read 1 operation will cause the cell to flip to 0; however, if only a single write 1 or a single read 1, or a read 1 which is not immediately applied after write 1 operation is performed, then the cell will not flip. [7] observed the existence of dynamic faults in the new embedded DRAMs based on defect injection and SPICE simulation. [33] observed the presence of dynamic faults in embedded caches of Pentium processor during a detailed analysis of the DPM screening results of a large number of test patterns. [32] showed the importance of dynamic faults for new SRAM technologies by analyzing DMP screening results of Intel and STMicroelectronics products, and concluded that current and future SRAMs need to consider dynamic faults testability or leave substantial DPM on the table.

The tests currently used in the industry have been designed to target static faults and therefore may not detect/diagnose dynamic faults. This indicates the importance of dynamic faults for current and future memory technologies. The dynamic fault class, which has been ignored in the past, is now becoming important and has to be taken into consideration. This sets a new direction for further research on memory fault modeling. Items like the following need to be worked out:

- Establishing the complete fault space, the fault framework (based on technology, design and time constraints) and the fault models for dynamic faults.

- Validation based on defect injection and SPICE simulation.

- IFA in order to determine the occurrence probabilities and the importance of each introduced fault models, and provide better understanding of the underlying defects causing the dynamic faults.

11.2.2 Other fault modeling aspects

Another special property of memories is that they have signals with a very high fan out. Examples of such signals are bit lines, word lines and address decoder pre-select lines. As the memories grow in size and operate on faster speeds, the lines carrying those signals will have, in addition to a high load, also a high parasitic capacitance. This all makes them very sensitive for delay and timing related faults because of their capacitive coupling with other signals, power and ground lines. Moreover, the significance of the resistance opens is considered to increase in recent and future technologies due to the presence of many, long interconnections and the growing number of metal layers and vias. As the partial resistive opens behave as delay and time related faults, these faults will become more important in the deep-submicron technologies.

Another aspect that has to be taken into consideration for the deep submicron technologies is the soft errors. The increased operation speed and noise margin reduction that accompany the technological scaling, are reducing continuously the reliability of new technologies memories due to the various internal sources of noise. This process in now approaching a point where it will be infeasible to produce memories that are free from these effects. The nanometer ICs are becoming so sensitive that even sea level radiation will introduce unacceptable soft errors [62]. Designing soft error tolerant circuits is the only way to follow the technological scaling. Among the most efficient techniques are error detecting and error correcting codes. This will not only detect and correct the soft errors and the new failure, but also compensate for the possible incomplete test/diagnosis fault coverage.

Other considerations for fault modeling for new technologies are (but not limited to):

- Transistor Short channel effect: lowering the threshold voltage may make the drain leakage contribution significant.

- Cross talk effect and noise from power lines.

- The impact of process variation on the speed failures.

Research on the above topic will be a source of new fault models. This will allow for dealing with the new defects; hence the development of new optimal high coverage test and diagnosis algorithms. Therefore reducing the DMP level and enhancing a high repair capabilities. The greater the fault detection and localization coverage, the higher the repair efficiency; hence the higher obtained yield.

Fault modeling can be done in different ways:

- Mathematically: Although this method is not time consuming and requires no layout/design information, it may lead to incorrect modeling. This method can be used to determine the framework of all possible faults, but tells nothing about the importance of a certain fault as compared with the other one.

- Functional analysis of DPM screening results of test patterns primitive. This method requires a good understanding of the failed/passed patterns. Based on a detail analysis of such patterns primitives, one can exactly give the sequence of operations for which the memory fails (do not fail); and therefore the fault model.

- Defect injection and SPICE simulation. This is the most used method in our days although it is time consuming and expensive. However, it provides a better reliable and trustworthy fault models than the previous two methods. The accuracy of the fault model depends on the accuracy of the simulation model and on the accuracy of the model used for the injected defect.

- IFA and simulation. This is the most accurate way for fault modeling since it uses extracted parameters from the real layout of the chip. However, it is the most time consuming and the most expensive way.

11.3 Test algorithm design

Memory tests and fault detection have experienced a long evolutionary process. The early tests (typically before the 1980's) can be classified as the *Ad-Hoc* tests because of the absence of formal fault models and proofs. Tests as Scan, Galpat and Walking 1/0 [11] belong to this class. They have further the property that for a given fault coverage, the test time was very long (excluded Scan), typically of order of $O(n^2)$, which made them very uneconomical for larger memories.

After the introduction of fault models during the early of 1980's, march tests became the dominant type of tests. The advantages of march tests lay in two facts. First, the fault coverage of the considered/known models could be mathematically proven, although one could not have any idea about the correlation between the models and the defects in the real chips. Second, the test time for march tests were usually linear with the size of the memory, which make them acceptable form industrial point of view. Some well known march tests that have been shown to be efficient are: Mats+ [60], March C- [49], PMOVI [39], IFA 13n [21], etc. As new fault models have been introduced end 1990's, based on defect injection and SPICE simulation, other new march tests have been developed to deal with them. Examples of such tests are March SR [24] and March SS [30].

Conventional memory test algorithms are basically designed to detect static functional faults (that are likely to occur) in order to determine if the chip is defective or not; in other words, they are pass/fail tests for static faults. As it have been shown in the previous section, the importance of dynamic faults increases with the new memory technologies. In addition, the shrinking technology will be a source of previously unknown defects/faults. The traditional tests are becoming thus insufficient/ inadequate for the today's and the future high speed memories. Therefore, new appropriate test algorithms have to be developed. On the other hand, as the memories become to represent the significant part of the SoC and dominant the overall yield, memory fault diagnosis becomes very important. Diagnosis techniques play a key role during the rapid development of semiconductor memories for catching design and/or manufacturing errors and failures; hence improving the yield. Although diagnosis has been widely used for memories, it is considered an expensive process due to long test times and complex fault/failure analysis procedure. Efficient diagnosis algorithms will benefit the industry and will play a more important role in the future as the SoC market grows.

Considering the current situation in test algorithm design and today's industry needs, can be concluded that new test/diagnosis algorithms still need to be developed; such algorithms have to take into consideration the following practical issues:

- Optimality in terms of time complexity in order to reduce the overall test time.

- Regularity and symmetric structure as possible so that the self-test circuit can use minimum of silicon area.

- High defect coverage and diagnosis capability in order to increase the repair capabilities and the overall yield.

- Appropriate stress combinations (voltage, temperature, timing, etc) that facilitate the detection of marginal faults.

11.4 Built-in-self test (BIST)

It is difficult to test an embedded memory simply by applying test patterns directly to the chip's I/O pins, because the embedded memory's address, data, and control signals are usually not directly accessible through the I/O pins. The basic philosophy behind the BIST technique is: "let the hardware test itself"; i.e. enhance the functionality of the memory to facilitate self-test. Large (and expensive) external tests cannot provide the needed test stimulus to enable high speed nor high quality tests [57]. BIST is therefore the only practical and cost-effective solution for embedded SoC memories.

BIST engine, no matter what kind, can use pseudo-random or deterministic patterns. A Pseudo-random pattern is basically very helpful to test logic circuits. A memory, however, has regular structure and requires application of regular test patterns. In the early days of BIST, it was not unusual to see pseudo-random techniques applied to memory [19], however these days this approach is hardly used due to its low fault coverage [4]. BISTs based on pseudo-random patterns are utilized on an occasional basis in the characterization of a design. Their design is mainly based on a linear feedback shift register (LFSR), which employs series of latches and XOR gates to implement a certain primitive polynomial [88].

BISTs based on deterministic patterns are dominant for testing memories today. Deterministic patterns means that patterns are generated according to specified predetermined values (e.g., march tests). For the design of such BISTs, two techniques are mainly used: state machines and micro-codes.

A state machine BIST can generate a single simple pattern or complex suite of patterns [94]. This BIST is generally used in the industry to generate a single pattern (e.g., a single march test). However, a better memory test solution requires a suite of patterns; this makes the design of the state machine complex. A state machine

BIST, as the name indicates, can exist in a number of states, which are group of latches from very few to several hundreds [14]. The major limitation of such BIST lays in its quite restricted flexibility. Modifying the patterns require changing the BIST design.

Micro-code BIST is a programmable BIST [44, 79], and therefore does not have limited flexibility. As new technologies are introducing faults that were previously unknown, new fault models can become evident during the fabrication. The BIST should be thus modifiable to include new patterns covering the new faults. The micro-code BIST is the most flexible of all self-test structures. The memory patterns can be easy modified to assist in the characterization of the new memory designs. In addition, programmable BIST can be used in both manufacturing and in a system environment. Due to the flexibility property, different patterns can be utilized depending on the applications.

For BIST design, in addition to minimizing the performance penalty introduced for normal memory operations, an important additional criterion is to minimize the area and the pin overhead. Embedded memories are usually of different widths, and are much smaller than stand-alone memories, resulting in high BIST overhead. BIST technology combines several different areas; e.g., fault modeling and test design. As the memories are increasing in size and as SoCs are including several memories of different sizes, with different access protocols and timing, the BIST technology is facing several practical issues like:

- Minimizing BIST overhead in both silicon area and routing.

- Selecting the proper number of BIST controllers to be used (i.e., choosing the proper clustering for multiple arrays for BIST controller hardware sharing).

- Supporting diagnosis capabilities.

- Fulfilling the power budget constraints.

- Supporting different kind memories (single-port, multi-port).

11.5 Built-in-self-repair (BISR)

As the complexity and the size of the embedded memories keep increasing, several challenges appear in the scene of memory repair in order to improve the overall yield. Using redundancy or spare elements is one of the most known used ways to improve the yield.

The most traditional manner for performing memory repair is using external test and repair. It starts first with applying memory test algorithms (can be on chip (BIST) or by ATE) and collecting the response in order to build the bit-map to be stored in a large capture memory on the ATE via limited I/O bandwidth; of course a high quality diagnosis algorithms have a critical role in this step. The off-chip failure bit-map used to record faults requires a large memory. The ATE software then uses the failed bit-map to determine the best way to allocate redundant elements to replace the defective elements and to generate the reconfiguration data. An optimal repair algorithm has been shown to be NP hard [20] and therefore requires a long execution time. The generated reconfiguration data is thereafter used for hard repair (laser/ electrical fusing). The fusing equipments will program the memory by blowing the fuses corresponding to the defective memory cells. The repaired memory has finally to be retested in order to ensure that the repair was successful.

The limitations of external repair lay in different factors. It relies on the extensive ATE, which makes the test cost of the chip very high (about 40% of the overall manufacturing cost of the semiconductor chip [101]). Since the external method relies on general-redundancy allocation algorithms, it also has limited repair efficiency. In addition, it only provides a limited I/O bandwidth for sending the large failed bit-map from the embedded memories to ATE. Furthermore, laser repair is often very expensive and some times continuous periodic field repair is desired [102].

To deal with the above limitations and reduce the overall manufacturing cost, the memory has to be made self-repairable. This will be achieved by expanding the embedded test resources even further to include a storage repair data and soft configuration mechanisms. In other words, BIST resources for future embedded memories will require to move beyond fault detection to include failed bit-map diagnosis, redundancy analysis and self repair. Once the failed bit-map is generated, based on diagnosis algorithm, the repair efficiency depends mainly on the type and the amount of redundancy, and on the allocation algorithms. The most common types of redundancy are: row redundancy, column redundancy, or block redundancy. Many allocation algorithms for BISR solutions have been proposed where the failure information does not need to be stored off-chip. Generally speaking, they are simple solutions.

In [18] a self-repairing structure for high hierarchical ultra-large memory chips is introduced. The repair scheme uses spare rows with the memory blocks at the lowest level of hierarchy and block redundancy at the top level. Although global redundancy can repair a great variety of faults, because of its greater inherent flexibility, it suffers form a higher area overhead.

A BISR for high density SRAMs using only spare columns and a greedy algorithm for allocation is proposed in [42]. The repair of defective circuits occurs autonomously without external stimulus (e.g., laser repair) by mapping redundancy columns via multiplexors to functionally replace the defective cells. The repair takes place immediately once a defective cell is found. Therefore no failure information is needed to be stored.

In [10] a solution is proposed based on combination of spare rows and columns. The scheme is based on divide-and-conquer strategy, where the memory is partitioned into several small identical segments; each segment is repaired independent of the others. However, the technique limits the numbers of spares to only one spare row pair and one spare column per each memory block in order to make the hardware realization practical and feasible; e.g., used failure bit-map stores only one row and one column (simple). Its extensibility to more complicated spare structures is limited.

A two-dimensional (i.e., spare rows and spare columns) repair allocation algorithm using a simple heuristics based on a finite state machine is presented in [61]. The scheme requires the storage (on chip) of only the final addresses to be repaired and the repair information, which is supplied to the laser repair equipment as direct programming data. The evaluation of the technique with up to five spare rows and five spare columns reveals that the scheme does not always guarantee 100% reparability of the memory, and therefore the allocation is not always possible.

The BISR analyzer suggested in [41] is based on an exhaustive search of all possible repair solutions for embedded DRAMs; the method therefore guarantee 100% detection ability of the repairable chips. The analyzer uses m spare rows and n spare columns per bock (where $m \geq 2$ and $n \geq 2$) and search all possible solution $C(n+m,n) = (n+m)!/(m!n!)$. The scheme (called Comprehensive Real Time Exhaustive Search Test and Analysis, CRESTA) provides plural sub-analyzers; each tries a repair solution concurrently in a predetermined different order using the m spare rows and n spare column. If multiple solutions found, the analyzer chooses the solution with the minimum number of used spares. Although CRESTA improves the analysis speed drastically, requires no external failure bit, reduces test time (due to at speed testing), and provide a repair solution if it exist, it suffers form practical limitations. Its implementation is mainly based on CAM (Content Addressable Memory). The size of the required CAM increases exponentially with the number of spare rows and columns $(m+n)$; i.e., $(n+m)*C(m+n,n)$ cells. This leads to infeasible hardware and time complexity for larger values of m and n.

All known repair algorithms are not optimal and practically have restricted applications, for smaller number of spare elements only. New repair schemes for arbitrary number of spares yet guaranteeing the solution have to be found while considering practical issues like:

- Dealing with complicated memory structures. Generally, memory is divided into different blocks with different number of spare rows and columns [82]. Each block can be partitioned into the row and column dimensions, where the number of partitions can be different. Further, instead of considering single rows (columns), physical adjacent rows (columns) can also be grouped into groups in order to facilitate the spare-group wise. Moreover, spare elements can be coupled to certain memory partition allowing that partition to use certain predefined spares for repair [76].

- Low hardware cost. For the reduction in time complexity and hardware overhead of the corresponding implementation, algorithms of redundancy analysis should be designed as simple as possible.

- Test time reduction. This can be achieved by using BIST and BISR at speed.

- On fly repair. This will prevent sending large failed bitmaps via limited I/O bandwidth to the ATE, and therefore reducing the test time and the overall cost.

- Applicability. The scheme should be with minor changes applicable to different types of memories, such as dual-port SRAMs.

- Repair methodologies. They determine the conditions under which redundancy analysis is determined and repair is performed [101].

11.6   Putting all together

To generate a high quality test strategy for new (embedded) memory technology, a through procedure must be pursued. First the memory design (with its cell, pre-charge, sense amplifier, etc.) has to be well understood. The circuits need to be investigated not only in the way they are expected to operate, but also in the way each of the circuits operates in the presence of various defects. These defective and faulty operations need to be mapped into fault models. Once the memory design is understood and the proper fault models are generated, the best test patterns can be developed. Since no single test can achieve an acceptable DPM level, a suite of test patterns needs to be used. Understanding the design, fault models and tests is required in order to prevent shipping defective parts to customers. Redundancy and repair goes beyond that and are required to guarantee adequate yield on the vast majority of memories. The design, fault modeling and test design have to be revisited in light of redundancy. Redundancy algorithms need to be generated to allocate each redundancy dimension to the appropriate fails, whereby maximizing the yield. Once the design, fault models, test patterns, and redundancy algorithms are understood, the correct built-in-self testing scheme can be designed (using e.g., a micro-code) while achieving a very low DPM level and a very high overall yield.

Bibliography

[1] D.R. Aadsen, et. al. Automated BIST for Regular Structures Embedded in ASIC Devices. *AT&T Technical Journal*, 69(3):97–105, 1990.

[2] M.S. Abadir and J.K. Reghbati. Functional Testing of Semiconductor Random Access Memories. *ACM Computer Surveys*, 15(3):175–198, 1983.

[3] M. Abramovici, M.A. Breuer and A.D. Friedman. Digital Systems Testing and Testable Design. New York, 1990.

[4] D. Adams. *High Performance Memory Testing*. Kluwer Academic Publishers, MA, USA, 2003.

[5] R.D. Adams and E.S. Cooley. Analysis of a Deceptive Read Destructive Memory Fault Model and Recommended Testing. *Proc. of IEEE North Atlantic Test Workshop*, pages 27–32, 1996.

[6] Z. Al Ars and A.J. van de Goor. Impact of Memory Cell Array Bridges on the Faulty Behavior in Embedded DRAMs. *Proc. of IEEE Asian Test Symposium*, pages 282–289, 2000.

[7] Z. Al Ars and A.J. van de Goor. Static and Dynamic Behavior of Memory Cell Array Opens and Shorts in Embedded DRAMs. *Proc. of European Design and Test Conference*, pages 401–406, 2001.

[8] V.C. Alves, M. Nocolaidis, P. Lestrat and B. Courtois. Built-in self-test for multi-port memories. *Proc. of IEEE Int. Conference on Computer Aided Design (ICCAD)*, pages 248–251, 1991.

[9] K.G. Ashar. *Magnetic Disk Drive Technology: heads, media, Chanel, interface, and application*. IEEE Press, New York, 1997.

[10] D.K. Bhavsar. An Algorithm for Row-Column Self Repair of RAMs and Its Implementation in the Alpha 21264. *Proc. of the IEEE Int. Test Conference*, pages 311–317, 1999.

[11] M.A. Breuer and A. D Friedman. *Diagnosis and Reliable Design of Digital Systems*. Computer Science Press, 1976.

[12] J.R. Brown. Pattern Sensitivity in MOS memories. *Digest Symposium on Testing to Integrate Semiconductor Memories into Computer Mainframes*, pages 33–46, 1972.

[13] M.L. Bushnell and V.D. Agrawal. *Essentials of Electronic Testing*. Kluwer Academic Publishers, MA, USA, 2000.

[14] P. Camurati, et.al. Industrial BIST for Embedded RAMs. *IEEE Design & Test of Computers*, 12(3):86–95, 1995.

[15] K. Chakraborty and P. Mazumder. A Programmable Boundary Scan Technique for Board-Level, Parallel Functional Duplex March Testing for Word-Oriented Multi-port Static RAMs. *Proc. of European Design and Test Conference*, pages 330–334, 1997.

[16] K. Chakraborty and P. Mazumder. New March Tests for Multi-port RAM Devices. *Journal of Electronic Testing: Theory and Application*, 16(4):389–395, 2000.

[17] P. Y. Chee, P. C. Liu, and L. Siek. High-Speed Hybrid Current-Mode Sense Amplifier for CMOS SRAMs. *Electronics Letters*, 28(9):871–873, 1992.

[18] T. Chen and G. Sunada. Design Self-Testing and Self-Repairing Structure for High Hierarchical Ultra-Large Capacity Memory Chips. *IEEE Transactions on VLSI*, 1(2):88–97, June 1993.

[19] R. David, A. Fuentes and B Courtois. Random Patterns Testing Versus Deterministic Testing of RAMs. *IEEE Transactions on Computers*, 38(5):637–650, 1989.

[20] J.R. Day. A Fault-Driven Comprehensive Redundancy Algorithm. *IEEE Design & Test of Computers*, 2(2):169–173, June 1985.

[21] R. Dekker F. Beenker, and H. Thijssen. A Realistic Fault Models and Test Algorithms for Static random Access Memories. *IEEE Transactions on Computer-Aided Design*, 9(6):567–572, 1990.

[22] S.W. Director, W. Maly, and A.J. Storjwas. *VLSI Design for Manufacturing Yield Enhancement*. Kluwer Academic Publishers, 1990.

[23] M. J. Forsell. Are Multi-port Memories Physically Feasible. *Computer Architecture News*, 22(5):3–10, 1994.

[24] S. Hamdioui A.J. van de Goor. Experimental Analysis of Spot Defects in SRAMs: Realistic Fault Models and Tests. *Proc. of IEEE Asian Test Symposium*, pages 131–138, 2000.

[25] S. Hamdioui, A.J. van de Goor, D. Eastwick, and M. Rodgers. Realistic Fault Models and Test Procedure for Multi-Port Memories. *Proc. of IEEE Int. Workshop on Memory Technology, Design and Testing*, pages 65–72, 2001.

[26] S. Hamdioui, A.J. van de Goor, M. Rodgers, and D. Eastwick. March Tests for Realistic Faults in Two-port Memories. *Proc. of IEEE Int. Workshop on Memory Technology, Design and Testing*, pages 73–78, 2000.

[27] S. Hamdioui and A.J. van de Goor. Consequences of Port Restrictions on Testing Two-Port Memories. *Proc. of the IEEE Int. Test Conference*, pages 63–72, 1998.

[28] S. Hamdioui and A.J. van de Goor. Port Interference Faults in Two-Port Memories. *Proc. of the IEEE Int. Test Conference*, pages 1001–1010, 1999.

[29] S. Hamdioui and A.J. van de Goor. Efficient Tests for Realistic Faults in Dual-Port Memories. *IEEE Transactions on Computers*, 51(5):460–473, 2002.

[30] S. Hamdioui and A.J. van de Goor. March SS: A Test for All Static Simple RAM Faults. *Proc. of IEEE Int. Workshop on Memory Technology, Design and Testing*, pages 95–100, 2002.

[31] S. Hamdioui and A.J. van de Goor. Through Testing Any Multi-Port Memory with Linear Tests. *IEEE Transactions on Computer-Aided Design*, 21(2):217–231, 2002.

[32] S. Hamdioui, et. al. Importance of Dynamic Faults for New SRAM Technologies. *Proc. of IEEE European Test Conference*, pages 29–34, 2003.

[33] S. Hamdioui, Z. Al-ars and A.J. van de Goor. Testing Static and Dynamic Faults in Random Access Memories. *Proc. of IEEE VLSI Test Symposium*, pages 395–400, 2002.

[34] C.G. Hawkins and J.M. Soden. Electrical Characteristics and Testing Considerations for Gate Oxide Shorts in CMOS ICs. *Proc. of the IEEE Int. Test Conference*, pages 544–555, 1985.

[35] M. Inoue, et. al. A New Test Evaluation Chip for Lower Cost Memory Tests. *IEEE Design & Test of Computers*, 10(1):15–19, March 1993.

[36] M. Izimikawa and M. Yamashina. A Current Direction Sense Technique for Multi-port Memories. *IEEE Journal of Solid State Circuits*, 31(4):546–551, 1996.

[37] A. Jee and F.J. Ferguson. An Inductive Fault Analysis Tool for CMOS VLSI Circuits. *Proc. of IEEE VLSI Test Symposium*, pages 92–98, 1993.

[38] F. Jensen and N.E. Peterson. *Burn-in*. John Wiley & Sons, UK, 1982.

[39] J.H. De Jonge and A.J. Smeulders. Moving Inversions Test Pattern is Thorough, Yet Speedy. *In Comp. Design*, pages 169–173, 1976.

[40] H. Kadota, et al. A Register File Structure for the High Speed Microprocessor. *IEEE Journal of Solid State Circuits*, SC(17):892–897, 1982.

[41] T. Kawagoe, et.al. A Built-in Self-Repair Analyzer (CRESTA) for embedded DRAMs. *Proc. of the IEEE Int. Test Conference*, pages 567–573, 2000.

[42] I. Kim, et. al. Built In Self Repair For Embedded High Density SRAMs. *Proc. of the IEEE Int. Test Conference*, pages 1112–1119, 1998.

[43] K. Kinoshita, and K. K. Saluja. Built-In Testing Using an On-Chip Compact Testing Scheme. *IEEE Transactions on Computers*, C-35(10):862–870, 1986.

[44] H. Koike, et al. A BIST Scheme Using Microprogram ROM for Large Capacity Memories. *Proc. of the IEEE Int. Test Conference*, pages 815–822, 1990.

[45] S.Y. Kuo and W.K. Fuchs H. Efficient Spare Allocation for Reconfigurable Arrays. *IEEE Design & Test of Computers*, 4(1):24–31, February 1987.

[46] P. Landsberg, C. Tretz and C. Zukowski. An Efficient Macromodel for Static CMOS Multi-Port Memories. *In Proc. of the IEEE 1993 Custom Integrated Circuits Conference*, pages 8.2.1–8.2.4, 1993.

[47] Y.K. Malaiya and S.H.H. Su. A New Fault Model and Testing Techniques for CMOS IC Defects. *Proc. of the IEEE Int. Test Conference*, pages 25–43, 1982.

[48] W. Maly. Modeling of Lithography Related Yield Losses for CAD of VLSI Circuits. *IEEE Transactions on Computer-Aided Design*, 4(3):166–177, 1985.

[49] M. Marinescu. Simple and Efficient Algorithms for Functional RAM Testing. *Proc. of the IEEE Int. Test Conference*, pages 236–239, 1982.

[50] T. Matsumura. An Efficient Test Method for Embedded Multi-Port RAM with BIST Circuitry. *Proc. of IEEE Int. Workshop on Memory Technology, Design and Testing*, pages 62–67, 1995.

[51] P.C. Maxwell, R.C. Aitken. Test Sets and Reject Rates: All Fault Coverages Are Not Created Equal. *IEEE Design and Test of Computers*, pages 42–51, March 1993.

[52] P. Mazumder. Parallel Testing of Parametric Faults in a Tree-Dimensional Dynamic Random-Access Memory. *IEEE Journal of Solid State Circuits*, 24(4):933–941, 1988.

[53] P. Mazumder and J. K. Patel. An Efficient Design of Embedded Memories and their Testability Analysis using Markov Chains. *IEEE Transactions on Computers*, C-38(3):394–407, 1989.

[54] P. Mazumder and K. Chakraborty. *Testing and Testable Design of High-Density Random Access Memories*. Kluwer Academic Press, Boston, 1996.

[55] A. Meixner and J. Banik. Weak Write Test Mode: An SRAM Cell Stability Design for Test Techniques. *Proc. of the IEEE Int. Test Conference*, pages 309–318, 1996.

[56] O. Minato, et.al. A Hi-CMOSII 8K×8K Bit Static RAM. *IEEE Journal of Solid State Circuits*, SC-23(5):793–798, 1982.

[57] R. Mookerjee. Segmentation: A Technique for Adapting High-Performance Logic ATE to Test High-Density, High-Speed SRAMs. *Proc. of IEEE Int. Workshop on Memory Technology, Design and Testing*, pages 120–124, 1993.

[58] B. Nadeau-Dostie, A. Sulburt, and V. K. Agrawal. Serial Interfacing for Embedded-memory Testing. *IEEE Design and Test of Computers*, pages 52–63, 1990.

[59] R. Nair. Efficient Test Algorithms for Testing Semiconductor Random Access Memories. *IEEE Transactions on Computers*, C-28(3):572–567, 1978.

[60] R. Nair. An Optimal Algorithm for Testing Stuck-at Faults Random Access Memories. *IEEE Transactions on Computers*, C-28(3):258–261, 1979.

[61] S. Nakahara, et.al. Built-In Self-Test for GHz Embedded SRAMs Using Flexible Patterns Generator and New Repair Algorithm. *Proc. of the IEEE Int. Test Conference*, pages 301–307, 1999.

[62] M. Nicolaidis. Design for Soft Error Robustness To Rescue Deep Submicron Scaling. *Proc. of the IEEE Int. Test Conference*, page 1140, 1998.

[63] M. Nicolaidis, V. C. Alves, and H. Bederr. Testing Complex Couplings in Multi-Port Memories. *IEEE Transactions on VLSI*, 3(1):59–71, March 1995.

[64] K.J. O'Connor. The Twin Port Memory Cell. *IEEE Journal of Solid State Circuits*, 22:712–720, October 1987.

[65] A. Offerman. Automatic Memory Test Verification and Generation. Master's thesis, Delft University of Technology, Faculty of Electrical Engineering, Mekelweg 4, 2628 CD, Delft, April 1995.

[66] C.A. Papachristou and N.B. Saghal. An Improved Method for Detecting Functional Faults in Random Access Memories. *IEEE Transactions on Computers*, C-34(2):110–116, 1985.

[67] B. Prince. *Semiconductor Memories, a Handbook of Design and Manufacturing and Application*. John Wiley & Sons, West Sussex, 1991.

[68] M. J. Raposa. Dual Port Static RAM Testing. *Proc. of the IEEE Int. Test Conference*, pages 362–368, 1988.

[69] I.M. Ratiu and H.B. Bakoglu. Pseudo-random Built-In-Self-Test Methodology and Implementation for the IBM RISC Systems/6000 Processor. *IBM Journal of Research and Development*, 34(1):78–84, 1990.

[70] R. Rodriguez-Montanes, E.M.J.G. Bruls, and J. Figueras. Bridging Defects Resistance Measurements in CMOS Process. *Proc. of the IEEE Int. Test Conference*, pages 892–899, 1992.

[71] K.K. Saluja, et al. Built-In-Self-Testing RAM: A Practical Alternative. *IEEE Design & Test*, 4(1):42–51, 1987.

[72] K. Sasaki, et al. A 15ns, 1Mbit CMOS SRAM. *IEEE Journal of Solid State Circuits*, 23(5):1067–1072, 1988.

[73] K. Sasaki, et al. A 9ns 1-Mbit CMOS SRAM. *IEEE Journal of Solid State Circuits*, 24:1219–1225, 1989.

[74] K. Sasaki, et al. A 7ns, 140mW, 1-Mbit CMOS SRAM with Current Sense Amplifier. *IEEE Journal of Solid State Circuits*, 27(11):1511–1518, 1992.

[75] I. Schanstra and A.J. van de Goor. Industrial evaluation of Stress Combinations for March Tests Applied to SRAMs. *Proc. of the IEEE Int. Test Conference*, pages 983–992, 1999.

[76] A. Sehgal, et. al. Yield Analysis for Repairable Embedded Memories. *Proc. of IEEE European Test Conference*, pages 35–40, 2003.

[77] Semiconductor Industry Association. *The National Roadmap for Semiconductors*. 181 Metro Drive, Ste. 450, San Jose, California, 1997, 1997.

[78] J. P. Shen. Inductive Fault Analysis of CMOS Integrated Circuits. *IEEE Design and Test of Computers*, pages 13–26, 1985.

[79] P.G. Shepard, et.al. Programmable Built-In Self-Test Method and Controller for Arrays. *USA patent Number 5.633.877*, 1997.

[80] R.W. Sherburne et al. A 32 bit NMOS Microprocessor with a Large Register File. *IEEE Journal of Solid State Circuits*, 19(5):682–689, October 1984.

[81] S. Shinagawa, et al. A MultiSpeed Digital Cross Connect Switching VLSI Using New Circuit Techniques for Dual Port RAMs. *IEEE Custom Integrated Circuits Conference*, pages 3.4.1–4, 1991.

[82] S. Shoukourian, V. Vardanian and Y. Zorian. An Approach for Evaluating of Redundancy Analysis Algorithms. *Proc. of IEEE Int. Workshop on Memory Technology, Design and Testing*, pages 51–55, 2001.

[83] Silicon Compiler Systems Corporation. Genesil System Compiler Library. 1, Blocks, 1988.

[84] J.M. Soden, et al. I_{DDQ} Testing: A review. *Journal of Electronic Testing: Theory and Application*, 3(4):291–304, 1992.

[85] M.R. Spica and B. Roeder. Yield Prediction with IFA Tools. *Intel Corporation*, 1999.

[86] A.K. Stevens. *Introduction to Component Testing*. Addison-Wesley, 1986.

[87] D.S. Suk and S.M. Reddy. A March Test for Functional Faults in Semiconductors Random-Access Memories. *IEEE Transactions on Computers*, C-30(12):982–985, 1981.

[88] A.J. van de Goor. *Testing Semiconductor Memories, Theory and Practice.* ComTex Publishing, Gouda, The Netherlands, 1998.

[89] A.J. van de Goor and C.A. Verruijt. An Overview of Deterministic Functional RAM Chip Testing. *ACM Computer Surveys*, 22(1):5–33, 1990.

[90] A.J. van de Goor and J. de Neef. Industrial Evaluation of DRAMs Tests. *In Proc. of Design Automation and Test in Europe*, pages 623–630, March 1999.

[91] A.J. van de Goor and J.E. Simonse. Defining SRAM resistive defects and their simulation stimuli. *Proc. of IEEE Asian Test Symposium*, pages 33–40, 1999.

[92] A.J. van de Goor and S. Hamdioui. Fault Models and Tests for Two-Port Memories. *Proc. of IEEE VLSI Test Symposium*, pages 401–410, April 1998.

[93] A.J. van de Goor and Z. Al-Ars. Functional Fault Models: A Formal Notation and Taxonomy. *Proc. of IEEE VLSI Test Symposium*, pages 281–289, 2000.

[94] J. van Sas, et.al. BIST For Embedded Static RAMs with Coverage Calculations. *Proc. of the IEEE Int. Test Conference*, pages 339–347, 1993.

[95] H. Veendrick. *Deep-Submicron CMOS ICs: From Basics to ASICs.* Kluwer Academic Publishers, ISBN 90-440-01116, second edition, 2000.

[96] H. Wang, P. C. Liu, and K. T. Lau. Low Power Dual-Port CMOS SRAM Macro Design. *Electronics Letters*, 32(15), July 1996.

[97] S. Wood, R. et al. A 5Gb/s 9-Port Application Specific SRAM with Built-In Self-Test. *Proc. of IEEE Int. Workshop on Memory Technology, Design and Testing*, pages 68–73, 1995.

[98] Y. Wu, and S. Gupta. Built-In Self Test for Multi-Port RAMs. *Proc. of IEEE Asian Test Symposium*, pages 398–403, 1997.

[99] H. Yokoyama, H. Tamamoto and X. Wen. Built-In Random Testing for Dual-Port RAMs. *Proc. of IEEE Int. Workshop on Memory Technology, Design and Testing*, pages 2–6, 1994.

[100] J. Zhao, et al. Detection of Inter-Port Faults in Multi-Port Static SRAMs. *Proc. of IEEE VLSI Test Symposium*, pages 297–302, 2000.

[101] Y. Zorian. Embedded-Memory Test and Repair: Infrastructure IP for SOC Yield. *Proc. of the IEEE Int. Test Conference*, pages 340–349, 2002.

[102] Y. Zorian, S. Dey and M. Rodgers. Test of Future System-on Chip. *Proc. of the IEEE Int. Test Conference*, pages 392–398, 2000.

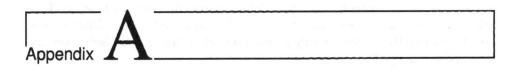

Simulation results for two-port SRAMs

A.1 Simulation results for opens

A.2 Simulation results for shorts

A.3 Simulation results for bridges

The minimal set of to be simulated spot defects in two-port memories consists of 18 opens, 6 shorts, and 25 bridges; see Table 5.1 and Table 5.2 in Chapter 5. All these defects have been simulated using the simulation model and methodology described in Chapter 4.

This appendix gives the simulation results, expressed in terms of fault primitives, for all 49 simulated defects. First, the simulation results for opens will be presented; then for shorts, and thereafter for bridges.

A.1   Simulation results for opens

Opens have been divided into opens within a cell (OCs) and opens at bit lines and word lines (OBs, OWs); see Section 5.1. The simulation has been done for all opens of Table 5.1. Each electrical faulty behavior is reported in terms of a fault primitive (FP); if a strong fault is sensitized, then the FP notation, introduced Chapter 3, is used to describe it. If a fault is only partially sensitized (i.e., a weak fault) then the fault is denoted as wF.

Table A.1 shows all simulation results for OCs; while Table A.2 presents the simulation results for OBs and OWs. The first column in each table gives the name of the open (see also Table 5.1); the second, the third and the fourth columns give the resistance regions[1] of the FP sensitized by the simulated open, and the derived complementary fault (if applicable), respectively. The latter will be the case when there is another defect (i.e, open) which has a complementary fault behavior as that of the simulated open (see Section 4.3). The fifth column gives the class of the sensitized FP; i.e., single-port faults (1PFs) divided into 1PFs involving a single-cell (1PF1s) and 1PFs involving two cells (1PF2s); and two-port faults (2PFs) divided into 2PFs involving a single-cell (2PF1s) and 2PFs involving two cells (2PF2s). It should be noted that in case the simulated open causes the gate of a certain transistor to be floating (for $R_{op} = \infty$), the simulation has been repeated for different initial gate voltages (by stepping between 0 and V_{cc} with a step size of $0.1V$) in order to get all possible fault effects. For example the open OC7 causes the gate of the pull down transistor at the true side to be floating for $R_{op} = \infty$ (see Figure 5.1); therefore during the simulation, the initial voltage of the floating gate (V_f) is examined in the range 0-V_{cc}; two fault behaviors have been observed (see Table A.1): $< 0/1/- >$ for $V_f \geq V_1^1$, and $< 1/0/- >$ for $V_f < V_1$. The tables clearly show that the sensitized fault is *strongly dependent* on the resistance value of the defect. They also show that the opens OC3 and OC4 cause 1PFs as well as 2PFs (depending on the resistance value of the defect), while the rest of the only opens cause 1PFs (1PF1s and/or 1PF2s).

---

[1]The exact values are design specific and Intel proprietary

Table A.1. Simulation results for OCs; $d=$ don't care value

Label	R_{op} region	Fault behavior	Compl. behavior	Class	Fault model
OC1	Region I	wF	wF	-	-
	Region II	$< 1_T/0/- >_v$	$< 0_T/1/- >_v$	1PF1	DRF
		wF	wF	-	-
OC2	Region I	wF	wF	-	-
	Region II	$< 1_T/0/- >_v$	$< 0_T/1/- >_v$	1PF1	DRF
		wF	wF	-	-
OC3	Region I	wF	wF	-	-
	Region II	$< 0r0 : 0r0/1/0 >_v$	$< 1r1 : 1r1/0/1 >_v$	2PF1	wDRDF&wDRDF
	Region III	$< 0r0 : 0r0/1/1 >_v$	$< 1r1 : 1r1/0/0 >_v$	2PF1	wRDF&wRDF
	Region IV	$< 0r0/1/0 >_v$	$< 1r1/0/1 >_v$	1PF1	DRDF
	Region V	$< 0r0/1/1 >_v$	$< 1r1/0/0_v$	1PF1	RDF
	Region VI	$< 0r0/1/1 >_v$	$< 1r1/0/0_v$	1PF1	RDF
		$< 0_T/1/- >_v$	$< 1_T/0/- >_v$	1PF1	DRF
OC4	Region I	wF	wF	-	-
	Region II	$< 0r0 : 0r0/1/0 >_v$	$< 1r1 : 1r1/0/1 >_v$	2PF1	wDRDF&wDRDF
	Region III	$< 0r0 : 0r0/1/1 >_v$	$< 1r1 : 1r1/0/0 >_v$	2PF1	wRDF&wRDF
	Region IV	$< 0r0/1/0 >_v$	$< 1r1/0/1 >_v$	1PF1	DRDF
	Region V	$< 0r0/1/1 >_v$	$< 1r1/0/0 >_v$	1PF1	RDF
	Region VI	$< 0r0/1/1 >_v$	$< 1r1/0/0 >_v$	1PF1	RDF
		$< 0_T/1/- >_v$	$< 1_T/0/- >_v$	1PF1	DRF
OC5	Region I	wF	wF	-	-
	Region II	wF	wF	-	-
		$< 1_T/0/- >_v$	$< 0_T/1/- >_v$	1PF1	DRF
	$R_{op} = \infty \ (V_f < V_1)$	wF	wF	-	-
	$R_{op} = \infty \ (V_f \geq V_1)$	wF	wF	-	-
		$< 1_T/0/- >_v$	$< 0_T/1/- >_v$	1PF1	DRF
OC6	Region I	$< 0w1/0/- >_v$	$< 1w0/1/- >_v$	1PF1	TF
	Region II	$< 0w1/0/- >_v$	$< 1w0/1/- >_v$	1PF1	TF
		$< 1w0/1/- >_v$	$< 1w0/0/- >_v$	1PF1	TF
	Region III	$< 1/0/- >_v$	$< 0/1/- >_v$	1PF1	SF
	$R_{op} = \infty; \ (V_f \geq V_1)$	$< 0/1/- >_v$	$< 1/0/- >_v$	1PF1	SF
	$R_{op} = \infty; \ (V_f < V_1)$	$< 1/0/- >_v$	$< 0/1/- >_v$	1PF1	SF
OC7	Region I	$< 0w1/0/- >_v$	$< 0w1/0/- >_v$	1PF1	TF
	Region II	$< 0r0/1/1 >_v$	$< 1r1/0/0 >_v$	1PF1	RDF
	$R_{op} = \infty; \ (V_f \geq V_1)$	$< 1/0/- >_v$	$< 0/1/- >_v$	1PF1	SF
	$R_{op} = \infty; \ (V_f < V_1)$	$< 0r0/1/1 >_v$	$< 1r1/0/0 >_v$	1PF1	RDF
OC8, OC9	Region I	$< 1w0/1/- >_v$	$< 0w1/0/- >$	1PF1	TF
		$< 0r0/0/? >_v$	$< 1r1/1/? >_v$	1PF1	RRF
OC10	Region I	$< 1w0/1/- >_v$	$< 0w1/0/- >_v$	1PF1	TF
		$< 0r0/0/? >_v$	$< 1r1/1/? >_v$	1PF1	RRF
OC11	Region I	$< 1w0/1/- >_v$	$< 0w1/0/- >_v$	1PF1	TF
		$< 0r0/0/? >_v$	$< 1r1/1/? >_v$	1PF1	RRF
	$R_{op} = \infty; \ (V_f < V_1)$	$< 1w0/1/- >_v$	$< 0w1/0/- >_v$	1PF1	TF
		$< 0r0/0/? >_v$	$< 1r1/1/? >_v$	1PF1	RRF
	$R_{op} = \infty; \ (V_f \geq V_1)$	$< dw0; 1/0/- >_{a,v}$	$< dw1; 0/1/- >_{a,v}$	1PF2	CFds
		$< 0; 1r1/1/? >_{a,v}$	$< 1; 0r0/0/? >_{a,v}$	1PF2	CFrr
OC12	Region I	wF		-	-
		$< 1_T/?/- >_v$		1PF1	-
		$< 0_T/?/- >_v$		1PF1	-
OC13	Region I	wF		-	-
	Region II	$< 0r0/?/? >_v$		1PF1	URF
		$< 1r1/?/? >_v$		1PF1	URF
		$< 1_T/?/- >_v$		1PF1	DRF
		$< 0_T/?/- >_v$		1PF1	DRF
OC14	Region I	wF		-	-
		$< 1_T/?/- >_v$		1PF1	-
		$< 0_T/?/- >_v$		1PF1	-
OC15	Region I	wF		-	-
	Region II	$< 0r0/?/? >_v$		1PF1	URF
		$< 1r1/?/? >_v$		1PF1	URF
		$< 1_T/?/- >_v$		1PF1	DRF
		$< 0_T/?/- >_v$		1PF1	DRF

Table A.2. Simulation results for OBs and OWs

Label	R_{op} region	Fault behavior	Compl. behavior	Class	Fault model
OBw	Region I	$< 1w0/1/- >_v$	$< 0w1/0/- >_v$	1PF1	TF
OBr	Region I	wF	wF	-	-
	Region II	$< 0r0/0/? >_v$	$< 1r1/1/? >_v$	1PF1	RRF
OW	Region I	$< 0w1/0/- >_v$		1PF1	TF
		$< 1w0/1/- >_v$		1PF1	TF
	Region II	$\{< 1w0/1/-_v >, < 1w0/1/- >_v$ $< 1r1/1/? >_v, < 0r0/0/? >_v\}$		1PF1	NAF
	$R_{op} = \infty; (V_f \geq V_1)$	$< dw1; 0/1/- >_{a,v}$		1PF2	CFds
		$< dw0; 1/0/- >_{a,v}$		1PF2	CFds
		$< 1; 0r0/0/? >_v$		1PF2	CFrr
		$< 0; 1r1/1/? >_v$		1PF2	CFrr
	$R_{op} = \infty; (V_2 \leq V_f < V_1)$	$< 0w1/0/- >_v$		1PF1	TF
		$< 1w0/1/- >_v$		1PF1	TF
	$R_{op} = \infty; (V_f < V_2)$	$\{< 1w0/1/? >_v, < 1w0/1/- >_v$ $< 1r1/1/? >_v, < 0r0/0/? >_v\}$		1PF1	NAF

A.2   Simulation results for shorts

Shorts have been divided into shorts within a cell (SCs) and shorts at bit lines (SBs) and at word lines (SWs); see Table 5.2. Table A.3 shows the simulation results of these shorts. The table shows that the sensitized FP depends on the resistance value of a short. In addition, it shows that all shorts only cause 1PFs, except the short SC2; the latter causes 1PFs as well as 2PFs.

Table A.3. Simulation results for shorts

Label	R_{op} region	Fault behavior	Compl. behavior	Class	Fault model
SC1	Region I	$< 0/1/- >_v$	$< 1/0/- >_v$	1PF1	SF
	Region II	$< 1w0/1/- >_v$	$< 0w1/0/- >_v$	1PF1	TF
	Region III	wF	wF	-	-
SC2	Region I	$< 1/0/- >_v$	$< 0/1/- >_v$	1PF1	SF
	Region II	$< 1r1/0/0 >_v$	$< 0r0/1/1 >_v$	1PF1	RDF
	Region III	$< 1r1/0/1 >_v$	$< 0r0/1/0 >_v$	1PF1	DRDF
	Region IV	$< 1r1 : 1r1/0/0 >_v$	$< 0r0 : 0r0/1/1 >_v$	2PF1	wRDF&wRDF
	Region V	$< 1r1 : 1r1/0/1 >_v$	$< 0r0 : 0r0/1/0 >_v$	2PF1	wDRDF&wDRDF
	Region VI	wF	wF	-	-
SB1	Region I	$< 1w0/1/- >_v$		1PF1	TF
	Region II	wF		-	-
SB2	Region I	$< 0w1/0/- >_v$	$< 1w0/1/- >_v$	1PF1	TF
	Region II	$< 1r1/0/0 >_v$	$< 0r0/1/1 >_v$	1PF1	RDF
	Region III	$< 1r1/1/0 >_v$	$< 0r0/0/1 >_v$	1PF1	IRF
	Region IV	$< 1r1/1/? >_v$	$< 0r0/0/? >_v$	1PF1	RRF
SW1	Region I	$< dw1; 0/1/- >_{a,v}$		1PF2	CFds
		$< dw0; 1/0/- >_{a,v}$		1PF2	CFds
SW2	Region I	$\{< 1w0/1/- >_v, < 1w0/1/- >_v$ $< 1r1/1/? >_v, < 0r0/0/? >_v\}$		1PF1	NAF
	Region II	$< 0w1/0/- >_v$		1PF1	TF
		$< 1w0/1/- >_v$		1PF1	TF

A.3 Simulation results for bridges

Bridges have been divided into bridges within a cell (BCs), and bridges between cells (BCCs). The BCCs are further divided into BCCs in the same row (rBCCs), BCCs in the same column (cBCCs), and BCCs on the same diagonal (dBCCs).

Table A.4 shows the simulation results for BCs. It is interesting to note that the two bridges BC6 and BC7 only cause 2PFs.

Table A.4. Simulation results for BCs; d = don't care value

Label	R_{br} region	Fault behavior	Compl. behavior	Class	Fault model
BC1	Region I	$< dw0/?/- >_v$		1PF1	UWF
		$< dw1/?/- >_v$		1PF1	UWF
	Region II	wF		-	-
BC2	Region I	$< 0/1/- >_v$	$< 1/0/- >_v$	1PF1	SF
	Region II	$< dw0; 1/0/- >_{a,v}$	$< dw1; 0/1/- >_{a,v}$	1PF2	CFds
		$< 0; 1r1/1/0 >_{a,v}$	$< 1; 0r0/0/1 >_{a,v}$	1PF2	CFir
	Region III	$< 0; 1r1/1/0 >_{a,v}$	$< 1; 0r0/0/1 >_{a,v}$	1PF2	CFir
	Region IV	$< 0; 1r1/1/? >_{a,v}$	$< 1; 0r0/0/? >_{a,v}$	1PF2	CFrr
	Region V	wF	wF	-	-
BC3	Region I	$< 0/1/- >_v$	$< 1/0/- >_v$	1PF1	SF
	Region II	$< 0r0/0/1 >_v$	$< 1r1/1/0 >_v$	1PF1	IRF
		$< 0; 0r0/0/1 >_{a,v}$	$< 1; 1r1/1/0 >_{a,v}$	1PF2	CFir
		$< dw1; 1/0/- >_{a,v}$	$< dw0; 0/1/- >_{a,v}$	1PF2	CFds
	Region III	$< 0r0/0/1 >_v$	$< 1r1/1/0 >_v$	1PF1	IRF
		$< 0; 0r0/0/1 >_{a,v}$	$< 1; 1r1/1/0 >_{a,v}$	1PF2	CFir
	Region IV	$< 0r0/0/? >_v$	$< 1r1/1/? >_v$	1PF1	RRF
		$< 0; 0r0/0/? >_{a,v}$	$< 1; 1r1/1/? >_{a,v}$	1PF2	CFrr
	Region V	wF	wF	-	-
BC4	Region I	$< 1/0/- >_v$	$< 0/1/- >$	1PF1	SF
	Region II	wF	wF	-	-
BC5	Region I	$\{< 1w0/1/- >_v, < 1w0/1/- >_v$			NAF
		$< 1r1/1/? >_v, < 0r0/0/? >_v\}$		1PF1	
	Region II	$< 0r0/0/? >_v$		1PF1	RRF
		$< 1r1/1/? >_v$		1PF1	RRF
	Region III	wF		-	-
BC6	Region I	$< dw0 : 1r1/0/0 >_{a,v}$	$< dw1 : 0r0/1/1 >_{a,v}$	2PF2	wCFds&wRDF
	Region II	$< dw0 : 1r1/1/0 >_{a,v}$	$< dw1 : 0r0/0/1 >_{a,v}$	2PF2	wCFds&wIRF
	Region III	$< dw0 : 1r1/1/? >_{a,v}$	$< dw1 : 0r0/0/? >_{a,v}$	2PF2	wCFds&wRRF
	Region IV	wF		-	-
BC7	Region I	$< 0r0 : 0w1/0/- >_v$		2PF1	wRDF&wTF
		$< 1r1 : 1w0/1/- >_v$		2PF1	wRDF&wTF
		$< dw1 : 1r1/0/0 >_{a,v}$		2PF2	wCFds&wRDF
		$< dw0 : 0r0/1/1 >_{a,v}$		2PF2	wCFds&wRDF
	Region II	$< dw1 : 1r1/0/0 >_{a,v}$		2PF2	wCFds&wRDF
		$< dw0 : 0r0/1/1 >_{a,v}$		2PF2	wCFds&wRDF
	Region III	$< dw1 : 1r1/1/0 >_{a,v}$		2PF2	wCFds&wIRF
		$< dw0 : 0r0/0/1 >_{a,v}$		2PF2	wCFds&wIRF
	Region IV	$< dw1 : 1r1/1/? >_{a,v}$		2PF2	wCFds&wRRF
		$< dw0 : 0r0/0/? >_{a,v}$		2PF2	wCFds&wRRF
	Region V	wF		-	-
BC8	Region I	$< 1w0/1/- >_v$	$< 0w1/0/- >_v$	1PF1	TF
		$< 1r1/0/0 >_v$	$< 0r0/1/1 >_v$	1PF1	RDF
	Region II	$< 1r1/0/0 >_v$	$< 0r0/1/1 >_v$	1PF1	RDF
	Region III	$< 1r1/1/0 >_v$	$< 0r0/0/1 >_v$	1PF1	IRF
	Region IV	$< 1r1/1/? >_v$	$< 0r0/0/? >_v$	1PF1	RRF
	Region V	wF	wF	-	-
BC9	Region I	$< 1r1/0/0 >_v$	$< 0r0/1/1 >_v$	1PF1	RDF
	Region II	$< 1r1/1/0 >_v$	$< 0r0/0/1 >_v$	1PF1	IRF
	Region III	$< 1r1/1/? >_v$	$< 0r0/0/? >_v$	1PF1	RRF
BC10	Region I	$< 0w1/0/- >_v$		1PF1	TF
		$< 1w0/1/- >_v$		1PF1	TF

Table A.5 and Table A.6 show the simulation results for rBCCs, cBCCs and dBCCs. Note that all BCCs cause conventional SP faults as well as 2PFs, except cBCC4 and cBCC5. Note also that the FPs caused by cBCC1 and dBCC1 are the same, as well as the FPs caused by cBCC2 and dBCC2.

Table A.5. Simulation results for BCCs; d=don't care value

Label	R_{br} region	Fault behavior	Compl. behavior	Class	Fault model
rBCC1	Region I	$< 0; 1/0/- >_{a,v}$	$< 1; 0/1/- >_{a,v}$	1PF2	CFst
	Region II	$< 0; 1r1/0/0 >_{a,v}$	$< 1; 0r0/1/1 >_{a,v}$	1PF2	CFrd
		$< 0r0; 1/0/- >_{a,v}$	$< 1r1; 0/1/- >_{a,v}$	1PF2	CFds
	Region III	$< 0; 1r1 : 1r1/0/0 >_{a,v}$	$< 1; 0r0 : 0r0/1/1 >_{a,v}$	2PF2	wCFrd&wRDF
		$< 0r0 : 0r0; 1/0/- >_{a,v}$	$< 1r1 : 1r1; 0/1/- >_{a,v}$	2PF2	wCFds&wCFds
		$< dw0 : drd; 1/0/- >_{a,v}$	$< dw1 : drd; 0/1/- >_{a,v}$	2PF2	wCFds&wCFds
	Region IV	$< 0; 1r1 : 1r1/0/1 >_{a,v}$	$< 1; 0r0 : 0r0/1/0 >_{a,v}$	2PF2	wCFdrd&wDRDF
		$< 0r0 : 0r0; 1/0/- >_{a,v}$	$< 1r1 : 1r1; 0/1/- >_{a,v}$	2PF2	wCFds&wCFds
		$< dw0 : drd; 1/0/- >_{a,v}$	$< dw1 : drd; 0/1/- >_{a,v}$	2PF2	wCFds&wCFds
	Region V	wF	wF	-	-
rBCC2	Region I	$< 0; 0/1/- >_{a,v}$	$< 1; 1/0/- >_{a,v}$	1PF2	CFst
		$< 1; 1/0/- >_{a,v}$	$< 0; 0/1/- >_{a,v}$	1PF2	CFst
	Region II	$< 1; 1r1/0/0 >_{a,v}$	$< 0; 0r0/1/1 >_{a,v}$	1PF2	CFrd
		$< 0; 0r0/1/1 >_{a,v}$	$< 1; 1r1/0/0 >_{a,v}$	1PF2	CFrd
		$< 0r0; 0/1/- >_{a,v}$	$< 1r1; 1/0/- >_{a,v}$	1PF2	CFds
		$< 1r1; 1/0/- >_{a,v}$	$< 0r0; 0/1/- >_{a,v}$	1PF2	CFds
	Region III	$< 1; 1r1 : 1r1/0/0 >_{a,v}$	$< 0; 0r0 : 0r0/1/1 >_{a,v}$	2PF2	wCFrd&wRDF
		$< 0; 0r0 : 0r0/1/1 >_{a,v}$	$< 1; 1r1 : 1r1/0/0 >_{a,v}$	2PF2	wCFrd&wRDF
		$< 0r0 : 0r0; 0/1/- >_{a,v}$	$< 1r1 : 1r1; 1/0/- >_{a,v}$	2PF2	wCFds&wCFds
		$< 1r1 : 1r1; 1/0/- >_{a,v}$	$< 0r0 : 0r0; 0/1/- >_{a,v}$	2PF2	wCFds&wCFds
		$< dw0 : drd; 0/1/- >_{a,v}$	$< dw1 : drd; 1/0/- >_{a,v}$	2PF2	wCFds&wCFds
		$< dw1 : drd; 1/0/- >_{a,v}$	$< dw0 : drd; 0/1/- >_{a,v}$	2PF2	wCFds&wCFds
	Region IV	$< 1; 1r1 : 1r1/0/1 >_{a,v}$	$< 0; 0r0 : 0r0/1/0 >_{a,v}$	2PF2	wCFdrd&wDRDF
		$< 0; 0r0 : 0r0/1/0 >_{a,v}$	$< 1; 1r1 : 1r1/0/1 >_{a,v}$	2PF2	wCFdrd&wDRDF
		$< 0r0 : 0r0; 0/1/- >_{a,v}$	$< 1r1 : 1r1; 1/0/- >_{a,v}$	2PF2	wCFds&wCFds
		$< 1r1 : 1r1; 1/0/- >_{a,v}$	$< 0r0 : 0r0; 0/1/- >_{a,v}$	2PF2	wCFds&wCFds
		$< dw0 : drd; 0/1/- >_{a,v}$	$< dw1 : drd; 1/0/- >_{a,v}$	2PF2	wCFds&wCFds
		$< dw1 : drd; 1/0/- >_{a,v}$	$< dw0 : drd; 0/1/- >_{a,v}$	2PF2	wCFds&wCFds
	Region V	wF	wF	-	-
rBCC3	Region I	$< 0/1/- >_v$	$< 1/0/- >_v$	1PF1	SF
	Region II	$< 1w0/1/- >_v$	$< 0w1/0/- >_v$	1PF1	TF
		$< dw0; 1/0/- >$	$< dw1; 0/1/- >$	1PF2	CFds
		$< 0; 1r1/1/0 >_{a,v}$	$< 1; 0r0/0/1 >_{a,v}$	1PF2	CFir
	Region III	$< 1; 1w0/1/- >_{a,v}$	$< 0; 0w1/0/- >_{a,v}$	1PF2	CFtr
		$< dw0; 1/0/- >_{a,v}$	$< dw1; 0/1/- >_{a,v}$	1PF2	CFds
		$< 0; 1r1/1/0 >_{a,v}$	$< 1; 0r0/0/1 >_{a,v}$	1PF2	CFir
	Region IV	$< dw0; 1/0/- >_{a,v}$	$< dw1; 0/1/- >_{a,v}$	1PF2	CFds
		$< 0; 1r1/1/0 >_{a,v}$	$< 1; 0r0/0/1 >_{a,v}$	1PF2	CFir
	Region V	$< 0; 1r1/1/0 >_{a,v}$	$< 1; 0r0/0/1 >_{a,v}$	1PF2	CFir
	Region VI	$< 0; 1r1/1/? >_{a,v}$	$< 1; 0r0/0/? >_{a,v}$	1PF2	CFrr
		$< dw0 : drd; 1/0/- >_{a,v}$	$< dw1 : drd; 0/1/- >_{a,v}$	2PF2	wCFds&wCFds
	Region VII	wF	wF	-	-
rBCC4	Region I	$< 0/1/- >_v$	$< 1/0/- >_v$	1PF1	SF
	Region II	$< 1w0/1/- >_v$	$< 0w1/0/- >_v$	1PF1	TF
		$< dw1; 1/0/- >_{a,v}$	$< dw0; 0/1/- >_{a,v}$	1PF2	CFds
		$< 0; 0r0/0/1 >_{a,v}$	$< 1; 1r1/1/0 >_{a,v}$	1PF2	CFir
	Region III	$< 0; 1w0/1/- >_{a,v}$	$< 1; 0w1/0/- >_{a,v}$	1PF2	CFtr
		$< dw1; 1/0/- >_{a,v}$	$< dw0; 0/1/- >_{a,v}$	1PF2	CFds
		$< 0; 0r0/0/1 >_{a,v}$	$< 1; 1r1/1/0 >_{a,v}$	1PF2	CFir
	Region IV	$< dw1; 1/0/- >_{a,v}$	$< dw0; 0/1/- >_{a,v}$	1PF2	CFds
		$< 0; 0r0/0/1 >_{a,v}$	$< 1; 1r1/1/0 >_{a,v}$	1PF2	CFir
	Region V	$< 0; 0r0/0/1 >_{a,v}$	$< 1; 1r1/1/0 >_{a,v}$	1PF2	CFir
	Region VI	$< 0; 0r0/0/? >_{a,v}$	$< 1; 1r1/1/? >_{a,v}$	1PF2	CFrr
		$< dw1 : drd; 1/0/- >_{a,v}$	$< dw0 : drd; 0/1/- >_{a,v}$	2PF2	wCFds&wCFds
	Region VII	wF	wF	-	-

Table A.6. Simulation results for BCCs (continue)

Label	R_{br} region	Fault behavior	Compl. behavior	Class	Fault model
rBCC5	Region I	$< dw0; 1/0/ >_{a,v}$	$< dw1; 0/1/ - / >_{a,v}$	1PF2	CFds
	Region II	$< dw0 : drd; 1/0/0 >_{a,v}$	$< dw1 : drd; 0/1/ - >_{a,v}$	2PF2	wCFds&wCFds
	Region III	wF	wF	-	-
rBCC6	Region I	$< dw0 : drd; 1/0/ - >_{a,v}$	$< dw1 : drd; 0/1/ - >_{a,v}$	2PF2	wCFds&wCFds
		$< dw0 : 1r1/0/0 >_{a,v}$	$< dw1 : 0r0/1/0 >_{a,v}$	2PF2	wCFds&wRDF
	Region II	$< dw0 : 1r1/0/0 >_{a,v}$	$< dw1 : 0r0/1/0 >_{a,v}$	2PF2	wCFds&wRDF
	Region III	$< dw0 : 1r1/1/0 >_{a,v}$	$< dw1 : 0r0/0/1 >_{a,v}$	2PF2	wCFds&wIRF
	Region IV	$< dw0 : 1r1/1/? >_{a,v}$	$< dw1 : 0r0/0/? >_{a,v}$	2PF2	wCFds&wRRF
	Region V	wF	wF	-	-
rBCC7	Region I	$< dw0; 0/1/ >$		1PF2	CFds
		$< dw1; 1/0/ >$		1PF2	CFds
	Region II	$< dw0 : drd; 0/1/ - >$		2PF2	wCFds&wCFds
		$< dw1 : drd; 1/0/ - >$		2PF2	wCFds&wCFds
	Region III	wF		-	-
rBCC8	Region I	$< dw0 : drd; 1/0/ - >_{a,v}$		2PF2	wCFds&wCFds
		$< dw1 : drd; 1/0/ - >_{a,v}$		2PF2	wCFds&wCFds
		$< dw1 : 1r1/0/0 >_{a,v}$		2PF2	wCFds&wRDF
		$< dw0 : 0r0/1/1 >_{a,v}$		2PF2	wCFds&wRDF
	Region II	$< dw1 : 1r1/0/0 >_{a,v}$		2PF2	wCFds&wRDF
		$< dw0 : 0r0/1/1 >_{a,v}$		2PF2	wCFds&wRDF
	Region III	$< dw1 : 1r1/1/0 >_{a,v}$		2PF2	wCFds&wIRF
		$< dw0 : 0r0/0/1 >_{a,v}$		2PF2	wCFds&wIRF
	Region IV	$< dw1 : 1r1/1/? >_{a,v}$		2PF2	wCFds&wRRF
		$< dw0 : 0r0/0/? >_{a,v}$		2PF2	wCFds&wRRF
	Region V	wF		-	-
cBCC1	Region I	$< 0; 1/0/ - >_{i,j}$	$< 1; 0/1/ - >_{i,j}$	1PF2	CFst
	Region II	$< 0; 1r1/0/0 >_{a,v}$	$< 1; 0r0/1/1 >_{a,v}$	1PF2	CFrd
	Region III	$< 0; 1r1/0/1 >_{a,v}$	$< 1; 0r0/1/0 >_{a,v}$	1PF2	CFdrd
	Region IV	$< 0; 1r1 : 1r1/0/0 >_{a,v}$	$< 1; 0r0 : 0r0/1/1 >_{a,v}$	2PF2	wCFrd&wRDF
	Region V	$< 0; 1r1 : 1r1/0/1 >_{a,v}$	$< 1; 0r0 : 0r0/1/0 >_{a,v}$	2PF2	wCFdrd&wDRDF
	Region VI	wF	wF	-	
cBCC2	Region I	$< 0; 1/0/ - >_{a,v}$	$< 1; 0/1/ - >_{a,v}$	1PF2	CFst
		$< 1; 0/1/ - >_{a,v}$	$< 0; 1/0/ - >_{a,v}$	1PF2	CFst
	Region II	$< 1; 1r1/0/0 >_{a,v}$	$< 0; 0r0/1/1 >_{a,v}$	1PF2	CFrd
		$< 0; 0r0/1/1 >_{a,v}$	$< 1; 1r1/0/0 >_{a,v}$	1PF2	CFrd
	Region III	$< 1; 1r1/0/1 >_{a,v}$	$< 0; 0r0/1/0 >_{a,v}$	1PF2	CFdrd
		$< 0; 0r0/1/0 >_{a,v}$	$< 1; 1r1/0/1 >_{a,v}$	1PF2	CFdrd
	Region IV	$< 1; 1r1 : 1r1/0/0 >_{a,v}$	$< 0; 0r0 : 0r0/1/1 >_{a,v}$	2PF2	wCFrd&wRDF
		$< 0; 0r0 : 0r0/1/1 >_{a,v}$	$< 1; 1r1 : 1r1/0/0 >_{a,v}$	2PF2	wCFrd&wRDF
	Region V	$< 1; 1r1 : 1r1/0/1 >_{a,v}$	$< 0; 0r0 : 0r0/1/0 >_{a,v}$	2PF2	wCFdrd&wDRDF
		$< 0; 0r0 : 0r0/1/0 >_{a,v}$	$< 1; 1r1 : 1r1/0/1 >_{a,v}$	2PF2	wCFdrd&wDRDF
	Region VI	wF	wF	-	
cBCC3	Region I	$< 1/0/ - >_{v}$	$< 0/1/ - >_{v}$	1PF1	SF
	Region II	$< 1r1/0/0 >_{v}$	$< 0r0/1/1 >_{v}$	1PF1	RDF
	Region III	$< 1r1/0/1 >_{v}$	$< 0r0/1/0 >_{v}$	1PF1	DRDF
	Region IV	$< 1r1 : 1r1/0/0 >_{v}$	$< 0r0 : 0r0/1/1 >_{v}$	2PF1	wRDF&wRDF
	Region V	$< 1r1 : 1r1/0/1 >_{v}$	$< 0r0 : 0r0/1/0 >_{v}$	2PF1	wDRDF&wDRDF
	Region VI	wF	wF	-	-
cBCC4	Region I	$< 0w1/0/ - >_{v}$		1PF1	TF
		$< 1w0/1/ - >_{v}$		1PF1	TF
cBCC5	Region I	$< 0w1/0/ - >_{v}$		1PF1	TF
		$< 1w0/1/ - >_{v}$		1PF1	TF
dBCC1	Region I	$< 0; 1/0/ - >_{i,j}$	$< 1; 0/1/ - >_{i,j}$	1PF2	CFst
	Region II	$< 0; 1r1/0/ - >_{a,v}$	$< 1; 0r0/1/ - >_{a,v}$	1PF2	CFrd
	Region III	$< 0; 1r1/0/1 >_{a,v}$	$< 1; 0r0/1/0 >_{a,v}$	1PF2	CFdrd
	Region IV	$< 0; 1r1 : 1r1/0/0 >_{a,v}$	$< 1; 0r0 : 0r0/1/1 >_{a,v}$	2PF2	wCFrd&wRDF
	Region V	$< 0; 1r1 : 1r1/0/1 >_{a,v}$	$< 1; 0r0 : 0r0/1/0 >_{a,v}$	2PF2	wCFrd&wDRDF
	Region VI	wF	wF	-	
dBCC2	Region I	$< 0; 1/0/ - >_{a,v}$	$< 1; 0/1/ - >_{a,v}$	1PF2	CFst
		$< 1; 0/1/ - >_{a,v}$	$< 0; 1/0/ - >_{a,v}$	1PF2	CFst
	Region II	$< 1; 1r1/0/0 >_{a,v}$	$< 0; 0r0/1/1 >_{a,v}$	1PF2	CFrd
		$< 0; 0r0/1/1 >_{a,v}$	$< 1; 1r1/0/0 >_{a,v}$	1PF2	CFrd
	Region III	$< 1; 1r1/0/1 >_{a,v}$	$< 0; 0r0/1/0 >_{a,v}$	1PF2	CFdrd
		$< 0; 0r0/1/0 >_{a,v}$	$< 1; 1r1/0/1 >_{a,v}$	1PF2	CFdrd
	Region IV	$< 1; 1r1 : 1r1/0/0 >_{a,v}$	$< 0; 0r0 : 0r0/1/1 >_{a,v}$	2PF2	wCFrd&wRDF
		$< 0; 0r0 : 0r0/1/1 >_{a,v}$	$< 1; 1r1 : 1r1/0/0 >_{a,v}$	2PF2	wCFrd&wRDF
	Region V	$< 1; 1r1 : 1r1/0/1 >_{a,v}$	$< 0; 0r0 : 0r0/1/0 >_{a,v}$	2PF2	wCFdrd&wDRDF
		$< 0; 0r0 : 0r0/1/0 >_{a,v}$	$< 1; 1r1 : 1r1/0/1 >_{a,v}$	2PF2	wCFdrd&wDRDF
	Region VI	wF	wF	-	

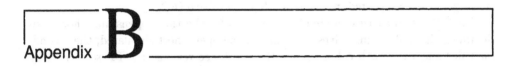

Simulation results for three-port SRAMs

B.1 Simulation results for opens and shorts

B.2 Simulation results for bridges

Based on the simulation results for two-port memories, the simulated defects have been divided into single-port fault defects (SFDs) and multi-port fault defects (MFDs). The SFDs are defects that can only cause single-port (SP) faults; they can not cause faults specific for multi-port (MP) memories. However, MFDs are defects that can cause SP faults as well as special faults for MP memories (e.g., two-port faults). That means that in order to analyze any MP memory with a number of ports $p > 2$, one needs to concentrate only on MFDs, since faults that can be caused by SFDs are SP faults, independent of the number of ports of the MP memory.

The minimal set of MFDs that has to be simulated for a 3P memory consists of 2 opens, 1 short, and 15 bridges; they are given in Table 8.1 and Table 8.2 in Chapter 8. All these defects have been simulated using the methodology described in Chapter 4. This appendix represents the simulation results for all MFDs in three-port memories. First, the simulation results for opens and shorts will be presented; and thereafter for bridges.

B.1   Simulation results for opens and shorts

The three MFDs listed in Table 8.1 have been simulated in the given order. The resistance of each open/short (R_{op}/R_{sh}) is examined in the range 0 to ∞. Each faulty behavior is reported in terms of a fault primitive (FP).

Table B.1 summarizes all simulation results for the three opens and shorts. In the table, the first column gives the name of the open/short simulated; the second, the third and the fourth columns give the resistance regions, the FP sensitized by the simulated MFD, and the derived complementary FP, respectively. The fifth column gives the class of the sensitized fault; i.e., single-port faults involving a single cell (1PF1s), two-port faults involving a single cell (2PF1s), or three-port faults involving a single cell (3PF1s). Note that the FP sensitized is strongly dependent on the resistance value of the defect, and that all simulated opens and shorts cause 1PFs, 2PFs, as well as 3PFs. In addition, the 3PF1s can be considered as extensions of the 2PFs.

Table B.1. Simulation results for opens and shorts causing multi-port faults

Label	R_{op}/R_{sh}	Fault behavior	Compl. behavior	Class	Fault model
OC3	Region I	$< 0r0/1/1 >$	$< 1r1/0/0 >$	1PF1	RDF
	Region II	$< 0r0/1/0 >$	$< 1r1/0/1 >$	1PF1	DRDF
	Region III	$< 0r0 : 0r0/1/1 >$	$< 1r1 : 1r1/0/0 >$	2PF1	wRDF&wRDF
	Region IV	$< 0r0 : 0r0/1/0 >$	$< 1r1 : 1r1/0/1 >$	2PF1	wDRDF&wDRDF
	Region V	$< 0r0 : 0r0 : 0r0/1/1 >$	$< 1r1 : 1r1 : 1r1/0/0 >$	3PF1	wRDF&wRDF&wRDF
	Region VI	$< 0r0 : 0r0 : 0r0/1/0 >$	$< 1r1 : 1r1 : 1r1/0/1 >$	3PF1	wDRDF&wDRDF&wDRDF
	Region VII	wF	wF	-	-
OC4	Region I	$< 0r0/1/1 >$	$< 1r1/0/0 >$	1PF1	RDF
	Region II	$< 0r0/1/0 >$	$< 1r1/0/1 >$	1PF1	DRDF
	Region III	$< 0r0 : 0r0/1/1 >$	$< 1r1 : 1r1/0/0 >$	2PF1	wRDF&wRDF
	Region IV	$< 0r0 : 0r0/1/0 >$	$< 1r1 : 1r1/0/1 >$	2PF1	wDRDF&wDRDF
	Region V	$< 0r0 : 0r0 : 0r0/1/1 >$	$< 1r1 : 1r1 : 1r1/0/0 >$	3PF1	wRDF&wRDF&wRDF
	Region VI	$< 0r0 : 0r0 : 0r0/1/0 >$	$< 1r1 : 1r1 : 1r1/0/1 >$	3PF1	wDRDF&wDRDF&wDRDF
	Region VII	wF	wF	-	-
SC2	Region I	$< 0/1/- >$	$< 1/0/- >$	1PF1	SF
	Region II	$< 1r1/0/0 >$	$< 0r0/1/1 >$	1PF1	RDF
	Region III	$< 1r1/0/1 >$	$< 0r0/1/0 >$	1PF1	DRDF
	Region IV	$< 1r1 : 1r1/0/0 >$	$< 0r0 : 0r0/1/1 >$	2PF1	wRDF&wRDF
	Region V	$< 1r1 : 1r1/0/1 >$	$< 0r0 : 0r0/1/0 >$	2PF1	wDRDF&wDRDF
	Region VI	$< 1r1 : 1r1 : 1r1/0/0 >$	$< 0r0 : 0r0 : 0r0/1/1 >$	3PF1	wRDF&wRDF&wRDF
	Region VII	$< 1r1 : 1r1 : 1r1/0/1 >$	$< 0r0 : 0r0 : 0r0/1/0 >$	3PF1	wDRDF&wDRDF&wDRDF
	Region VIII	wF	wF	-	-

B.2   Simulation results for bridges

In a similar way as for opens and shorts, the bridges for three-port memories have been simulated. Remember that the bridges have been divided into bridges within a cell (BCs) and bridges between cells (BCCs). The MPF bridges consist of 2 BCs and 13 BCCs; see Table 8.2. The 13 BCCs consist of 8 BCCs in the same row (rBCCs), 3 BCCs in the same column (cBCCs) and 2 BCCs on the same diagonal (dBCCs).

The simulation results for BCs are shown in Table B.2. The table shows clearly that all sensitized FPs are 2PFs (i.e, two-port faults); no 3PFs have been sensitized.

Therefore, one can conclude that in any MP memory, bridges BC6 and BC7 (i.e., bridges between bit lines belonging to the same column and different ports) can cause only 2PFs, irrespective of the number of ports the MP memory consist of.

Table B.2. Simulation results for BCs; d=don't care value

Label	R_{br} region	Fault behavior	Compl. behavior	Class	Fault model
BC6	Region I	$< dw0 : 1r1/0/0 >_{a,v}$	$< dw1 : 0r0/1/1 >_{a,v}$	2PF2	wCFds&wRDF
	Region II	$< dw0 : 1r1/1/0 >_{a,v}$	$< dw1 : 0r0/0/1 >_{a,v}$	2PF2	wCFds&wIRF
	Region III	$< dw0 : 1r1/1/? >_{a,v}$	$< dw1 : 0r0/0/? >_{a,v}$	2PF2	wCFds&wRRF
	Region IV	wF	wF	-	-
BC7	Region I	$< 0r0 : 0w1/0/ >_v$		2PF1	wRDF&wTF
		$< 1r1 : 1w0/1/ >_v$		2PF1	wRDF&wTF
		$< dw1 : 1r1/0/0 >_{a,v}$		2PF2	wCFds&wRDF
		$< dw0 : 0r0/1/1 >_{a,v}$		2PF2	wCFds&wRDF
	Region II	$< dw1 : 1r1/0/0 >_{a,v}$		2PF2	wCFds&wRDF
		$< dw0 : 0r0/1/1 >_{a,v}$		2PF2	wCFds&wRDF
	Region III	$< dw1 : 1r1/1/0 >_{a,v}$		2PF2	wCFds&wIRF
		$< dw0 : 0r0/0/1 >_{a,v}$		2PF2	wCFds&wIRF
	Region IV	$< dw1 : 1r1/1/? >_{a,v}$		2PF2	wCFds&wRRF
		$< dw0 : 0r0/0/? >_{a,v}$		2PF2	wCFds&wRRF
	Region V	wF		-	-

Table B.3, Table B.4 and Table B.5 summarizes the simulation results for rBCCs, cBCCs and dBCCs. The tables clearly show that, depending on the resistance value of the defect, rBCCs, cBCCs and dBCCs can cause 1PFs, 2PFs as well as 3PFs. The results also show that the FPs caused by cBCC1 and dBCC1 are similar, as well as the FPs caused by cBCC2 and dBCC2.

Table B.3. Simulation results for rBCCs; d=don't care

Label	R_{br}	Fault behavior	Compl. behavior	Class	Fault model
rBCC1	I	$< 0; 1/0/- >_{a,v}$	$< 1; 0/1/- >_{a,v}$	1PF2	CFst
	II	$< 0; 1r1/0/0 >_{a,v}$	$< 1; 0r0/1/1 >_{a,v}$	1PF2	CFrd
		$< 0r0; 1/0/- >_{a,v}$	$< 1r1; 0/1/- >_{a,v}$	1PF2	CFds
	III	$< 0; 1r1 : 1r1/0/0 >_{a,v}$	$< 1; 0r0 : 0r0/1/1 >_{a,v}$	2PF2	wCFrd&wRDF
		$< 0r0 : 0r0; 1/0/- >_{a,v}$	$< 1r1 : 1r1; 0/1/- >_{a,v}$	2PF2	wCFds&wCFds
		$< dw0 : drd; 1/0/- >_{a,v}$	$< dw1 : drd; 0/1/- >_{a,v}$	2PF2	wCFds&wCFds
	IV	$< 0; 1r1 : 1r1/0/1 >_{a,v}$	$< 1; 0r0 : 0r0/1/0 >_{a,v}$	2PF2	wCFdrd&wDRDF
		$< 0r0 : 0r0; 1/0/- >_{a,v}$	$< 1r1 : 1r1; 0/1/- >_{a,v}$	2PF2	wCFds&wCFds
		$< dw0 : drd; 1/0/- >_{a,v}$	$< dw1 : drd; 0/1/- >_{a,v}$	2PF2	wCFds&wCFds
	V	$< 0; 1r1 : 1r1 : 1r1/0/0 >_{a,v}$	$< 1; 0r0 : 0r0 : 0r0/1/1 >_{a,v}$	3PF2	wCFrd&wRDF&wRDF
		$< 0r0 : 0r0 : 0r0; 1/0/- >_{a,v}$	$< 1r1 : 1r1 : 1r1; 0/1/- >_{a,v}$	2PF2	wCFds&wCFds&wCFds
		$< dw0 : drd : drd; 1/0/- >_{a,v}$	$< dw1 : drd : drd; 0/1/- >_{a,v}$	3PF2	wCFds&wCFds&wCFds
	VI	$< 0; 1r1 : 1r1 : 1r1/0/1 >_{a,v}$	$< 1; 0r0 : 0r0 : 0r0/1/0 >_{a,v}$	3PF2	wCFdrd&wDRDF&wDRDF
		$< 0r0 : 0r0 : 0r0; 1/0/- >_{a,v}$	$< 1r1 : 1r1 : 1r1; 0/1/- >_{a,v}$	2PF2	wCFds&wCFds&wCFds
		$< dw0 : drd : drd; 1/0/- >_{a,v}$	$< dw1 : drd : drd; 0/1/- >_{a,v}$	3PF2	wCFds&wCFds&wCFds
	VII	wF	wF	-	-
rBCC2	I	$< 0; 0/1/- >_{a,v}$	$< 1; 1/0/- >_{a,v}$	1PF2	CFst
		$< 1; 1/0/- >_{a,v}$	$< 0; 0/1/- >_{a,v}$	1PF2	CFst
	II	$< 1; 1r1/0/0 >_{a,v}$	$< 0; 0r0/1/1 >_{a,v}$	1PF2	CFrd
		$< 0; 0r0/1/1 >_{a,v}$	$< 1; 1r1/0/0 >_{a,v}$	1PF2	CFrd
		$< 0r0; 0/1/- >_{a,v}$	$< 1r1; 1/0/- >_{a,v}$	1PF2	CFds
		$< 1r1; 1/0/- >_{a,v}$	$< 0r0; 0/1/- >_{a,v}$	1PF2	CFds
	III	$< 1; 1r1 : 1r1/0/0 >_{a,v}$	$< 0; 0r0 : 0r0/1/1 >_{a,v}$	2PF2	wCFrd&wRDF
		$< 0; 0r0 : 0r0/1/1 >_{a,v}$	$< 1; 1r1 : 1r1/0/0 >_{a,v}$	2PF2	wCFrd&wRDF
		$< 0r0 : 0r0; 0/1/- >_{a,v}$	$< 1r1 : 1r1; 1/0/- >_{a,v}$	2PF2	CFds&wCFds
		$< 1r1 : 1r1; 1/0/- >_{a,v}$	$< 0r0 : 0r0; 0/1/- >_{a,v}$	2PF2	CFds&wCFds
		$< dw0 : drd; 0/1/- >_{a,v}$	$< dw1 : drd; 1/0/- >_{a,v}$	2PF2	wCFds&wCFds
		$< dw1 : drd; 1/0/- >_{a,v}$	$< dw0 : drd; 0/1/- >_{a,v}$	2PF2	wCFds&wCFds
	IV	$< 1; 1r1 : 1r1/0/1 >_{a,v}$	$< 0; 0r0 : 0r0/1/0 >_{a,v}$	2PF2	wCFdrd&wDRDF
		$< 0; 0r0 : 0r0/1/1 >_{a,v}$	$< 1; 1r1 : 1r1/0/0 >_{a,v}$	2PF2	wCFdrd&wDRDF
		$< 0r0 : 0r0; 0/1/- >_{a,v}$	$< 1r1 : 1r1; 1/0/- >_{a,v}$	2PF2	wCFds&wCFds
		$< 1r1 : 1r1; 1/0/- >_{a,v}$	$< 0r0 : 0r0; 0/1/- >_{a,v}$	2PF2	wCFds&wCFds
		$< dw0 : drd; 0/1/- >_{a,v}$	$< dw1 : drd; 1/0/- >_{a,v}$	2PF2	wCFds&wCFds
		$< dw1 : drd; 1/0/- >_{a,v}$	$< dw0 : drd; 0/1/- >_{a,v}$	2PF2	wCFds&wCFds
	V	$< 1; 1r1 : 1r1 : 1r1/0/0 >_{a,v}$	$< 0; 0r0 : 0r0 : 0r0/1/1 >_{a,v}$	3PF2	wCFrd&wRDF&wRDF
		$< 0; 0r0 : 0r0 : 0r0/1/1 >_{a,v}$	$< 1; 1r1 : 1r1 : 1r1/0/0 >_{a,v}$	3PF2	wCFrd&wRDF&wRDF
		$< 0r0 : 0r0 : 0r0; 0/1/- >_{a,v}$	$< 1r1 : 1r1 : 1r1; 1/0/- >_{a,v}$	3PF2	wCFds&wCFds&wCFds
		$< 1r1 : 1r1 : 1r1; 1/0/- >_{a,v}$	$< 0r0 : 0r0 : 0r0; 0/1/- >_{a,v}$	3PF2	wCFds&wCFds&wCFds
		$< dw0 : drd : drd; 0/1/- >_{a,v}$	$< dw1 : drd : drd; 1/0/- >_{a,v}$	3PF2	wCFds&wCFds&wCFds
		$< dw1 : drd : drd; 1/0/- >_{a,v}$	$< dw0 : drd : drd; 0/1/- >_{a,v}$	3PF2	wCFds&wCFds&wCFds
	VI	$< 1; 1r1 : 1r1 : 1r1/0/1 >_{a,v}$	$< 0; 0r0 : 0r0 : 0r0/1/0 >_{a,v}$	3PF2	wCFdrd&wDRDF&wDRDF
		$< 0; 0r0 : 0r0 : 0r0/1/1 >_{a,v}$	$< 1; 1r1 : 1r1 : 1r1/0/1 >_{a,v}$	3PF2	wCFdrd&wDRDF&wDRDF
		$< 0r0 : 0r0 : 0r0; 0/1/- >_{a,v}$	$< 1r1 : 1r1 : 1r1; 1/0/- >_{a,v}$	3PF2	wCFds&wCFds&wCFds
		$< 1r1 : 1r1 : 1r1; 1/0/- >_{a,v}$	$< 0r0 : 0r0 : 0r0; 0/1/- >_{a,v}$	3PF2	wCFds&wCFds&wCFds
		$< dw0 : drd : drd; 0/1/- >_{a,v}$	$< dw1 : drd : drd; 1/0/- >_{a,v}$	3PF2	wCFds&wCFds&wCFds
		$< dw1 : drd : drd; 1/0/- >_{a,v}$	$< dw0 : drd : drd; 0/1/- >_{a,v}$	3PF2	wCFds&wCFds&wCFds
	V	wF	wF	-	-
rBCC3	I	$< 0/1/- >_v$	$< 1/0/- >_v$	1PF1	SF
	II	$< 1w0/1/- >_v$	$< 0w1/0/- >_v$	1PF1	TF
		$< 0; 1r1/1/0 >_{a,v}$	$< 1; 0r0/0/1 >_{a,v}$	1PF2	CFir
		$< dw0; 1/0/- >_{a,v}$	$< dw1; 0/1/- >_{a,v}$	1PF2	CFds
	III	$< 1; 1w0/1/- >_{a,v}$	$< 0; 0w1/0/- >_{a,v}$	1PF1	CFtr
		$< 0; 1r1/1/0 >_{a,v}$	$< 1; 0r0/0/1 >_{a,v}$	1PF2	CFir
		$< dw0; 1/0/- >_{a,v}$	$< dw1; 0/1/- >_{a,v}$	1PF2	CFds
	IV	$< 0; 1r1/1/0 >_{a,v}$	$< 1; 0r0/0/1 >_{a,v}$	1PF2	CFir
		$< dw0; 1/0/- >_{a,v}$	$< dw1; 0/1/- >_{a,v}$	1PF2	CFds
	V	$< 0; 1r1/1/0 >_{a,v}$	$< 1; 0r0/0/1 >_{a,v}$	1PF2	CFir
	VI	$< 0; 1r1/1/? >_{a,v}$	$< 1; 0r0/0/? >_{a,v}$	1PF2	CFrr
		$< dw0 : drd; 1/0/- >_{a,v}$	$< dw1 : drd; 0/1/- >_{a,v}$	2PF2	wCFds&wCFds
	VII	$< dw0 : drd : drd; 1/0/- >_{a,v}$	$< dw1 : drd : drd; 0/1/- >_{a,v}$	3PF2	wCFds&wCFds&wCFds
	VIII	wF	wF	-	-

Table B.4. Simulation results for rBCCs (continue)

Label	R_{br}	Fault behavior	Compl. behavior	Class	Fault model
rBCC4	I	$< 0/1/- >_v$	$< 1/0/- >_v$	1PF1	SF
	II	$< 1w0/1/- >_v$	$< 0w1/0/- >_v$	1PF1	TF
		$< 0; 0r0/0/1 >_{a,v}$	$< 1; 1r1/1/0 >_{a,v}$	1PF2	CFir
		$< dw1; 1/0/- >_{a,v}$	$< dw0; 0/1/- >_{a,v}$	1PF2	CFds
	III	$< 0; 1w0/1/- >_{a,v}$	$< 1; 0w1/0/- >_{a,v}$	1PF2	CFtr
		$< 0; 0r0/0/1 >_{a,v}$	$< 1; 1r1/1/0 >_{a,v}$	1PF2	CFir
		$< dw1; 1/0/- >_{a,v}$	$< dw0; 0/1/- >_{a,v}$	1PF2	CFds
	VI	$< 0; 0r0/0/1 >_{a,v}$	$< 1; 1r1/1/0 >_{a,v}$	1PF2	CFir
		$< dw1; 1/0/- >_{a,v}$	$< dw0; 0/1/- >_{a,v}$	1PF2	CFds
	V	$< 0; 0r0/0/1 >_{a,v}$	$< 1; 1r1/1/0 >_{a,v}$	1PF2	CFir
	VI	$< 0; 0r0/0/? >_{a,v}$	$< 1; 1r1/1/? >_{a,v}$	1PF2	CFrr
		$< dw1 : drd; 1/0/- >_{a,v}$	$< dw0 : drd; 0/1/- >_{a,v}$	2PF2	wCFds&wCFds
	VII	$< dw1 : drd : drd; 1/0/- >_{a,v}$	$< dw0 : drd : drd; 0/1/- >_{a,v}$	3PF2	wCFds&wCFds&wCFds
	VIII	wF	wF	-	-
rBCC5	I	$< dw0; 1/0/- >_{a,v}$	$< dw1; 0/1/- >_{a,v}$	1PF2	CFds
	II	$< dw0 : drd; 1/0/- >_{a,v}$	$< dw1 : drd; 0/1/- >_{a,v}$	2PF2	wCFds&wCFds
	III	$< dw0 : drd : drd; 1/0/- >_{a,v}$	$< dw1 : drd : drd; 0/1/- >_{a,v}$	3PF2	wCFds&wCFds&wCFds
	IV	wF	wF	-	-
rBCC6	I	$< dw0 : 1r1/0/0 >_{a,v}$	$< dw1 : 0r0/1/1 >_{a,v}$	2PF2	wCFds&wRDF
		$< dw0 : drd; 1/0/- >_{a,v}$	$< dw1 : drd; 0/1/- >_{a,v}$	2PF2	wCFds&wCFds
	II	$< dw0 : 1r1/0/0 >_{a,v}$	$< dw1 : 0r0/1/1 >_{a,v}$	2PF2	wCFds&wRDF
		$< dw0 : drd : drd; 1/0/- >_{a,v}$	$< dw1 : drd : drd; 0/1/- >_{a,v}$	3PF2	wCFds&CFds&CFds
	III	$< dw0 : 1r1/1/0 >_{a,v}$	$< dw1 : 0r0/0/1 >_{a,v}$	2PF2	wCFds&wIRF
	IV	$< dw0 : 1r1/1/? >_{a,v}$	$< dw1 : 0r0/0/? >_{a,v}$	2PF2	wCFds&wRRF
rBCC7	I	$< dw0; 1/0/- >_{a,v}$	$< dw1; 0/1/- >_{a,v}$	1PF2	wCFds
		$< dw1; 1/0/- >_{a,v}$	$< dw0; 0/1/- >_{a,v}$	1PF2	wCFds
	II	$< dw0 : drd; 1/0/- >_{a,v}$	$< dw1 : drd; 0/1/- >_{a,v}$	2PF2	wCFds&wCFds
		$< dw1 : drd; 1/0/- >_{a,v}$	$< dw0 : drd; 0/1/- >_{a,v}$	2PF2	wCFds&wCFds
	III	$< dw0 : drd : drd; 1/0/- >_{a,v}$	$< dw1 : drd : drd; 0/1/- >_{a,v}$	3PF2	wCFds&wCFds&wCFds
		$< dw1 : drd : drd; 1/0/- >_{a,v}$	$< dw0 : drd : drd; 0/1/- >_{a,v}$	3PF2	wCFds&wCFds&wCFds
	IV	wF	wF	-	-
rBCC8	I	$< dw0 : 0r0/1/1 >_{a,v}$	$< dw1 : 1r1/0/1 >_{a,v}$	2PF2	wCFds&wRDF
		$< dw1 : 1r1/0/0 >_{a,v}$	$< dw0 : 0r0/1/1 >_{a,v}$	2PF2	wCFds&wRDF
		$< dw0 : drd; 0/1/- >_{a,v}$	$< dw1 : drd; 1/0/- >_{a,v}$	2PF2	wCFds&wCFds
		$< dw1 : drd; 1/0/- >_{a,v}$	$< dw0 : drd; 0/1/- >_{a,v}$	2PF2	wCFds&wCFds
	II	$< dw0 : 0r0/1/1 >_{a,v}$	$< dw1 : 1r1/0/0 >_{a,v}$	2PF2	wCFds&wRDF
		$< dw1 : 1r1/0/0 >_{a,v}$	$< dw0 : 0r0/1/1 >_{a,v}$	2PF2	wCFds&wRDF
		$< dw0 : drd : drd; 0/1/- >_{a,v}$	$< dw1 : drd : drd; 1/0/- >_{a,v}$	3PF2	wCFds&wCFds&wCFds
		$< dw1 : drd : drd; 1/0/- >_{a,v}$	$< dw1 : drd : drd; 0/1/- >_{a,v}$	3PF2	wCFds&wCFds&wCFds
	III	$< dw0 : 0r0/1/0 >_{a,v}$	$< dw1 : 1r1/1/0 >_{a,v}$	2PF2	wCFds&wIRF
		$< dw1 : 1r1/1/0 >_{a,v}$	$< dw0 : 0r0/0/1 >_{a,v}$	2PF2	wCFds&wIRF
	IV	$< w0 : 0r0/0/? >_{a,v}$	$< w1 : 1r1/1/? >_{a,v}$	2PF2	wCFds&wRRF
		$< w1 : 1r1/1/? >_{a,v}$	$< w0 : 0r0/0/? >_{a,v}$	2PF2	wCFds&wRRF

Table B.5. Simulation results for cBCCs and dBCCs

Label	R_{br}	Fault behavior	Compl. behavior	Class	Fault model
cBCC1	I	$< 0;1/0/- >_{a,v}$	$< 1;0/1/- >_{a,v}$	1PF2	CFst
	II	$< 0;1r1/0/0 >_{a,v}$	$< 1;0r0/1/1 >_{a,v}$	1PF2	CFrd
	III	$< 0;1r1/0/1 >_{a,v}$	$< 1;0r0/1/0 >_{a,v}$	1PF2	CFdrd
	IV	$< 0;1r1 : 1r1/0/0 >_{a,v}$	$< 1;0r0 : 0r0/1/1 >_{a,v}$	2PF2	wCFrd&wRDF
	V	$< 0;1r1 : 1r1/0/1 >_{a,v}$	$< 1;0r0 : 0r0/1/0 >_{a,v}$	2PF2	wCFdrd&wDRDF
	VI	$< 0;1r1 : 1r1 : 1r1/0/0 >_{a,v}$	$< 1;0r0 : 0r0 : 0r0/1/1 >_{a,v}$	3PF2	wCFrd&wRDF&wRDF
	VII	$< 0;1r1 : 1r1 : 1r1/0/1 >_{a,v}$	$< 1;0r0 : 0r0 : 0r0/1/0 >_{a,v}$	3PF2	wCFdrd&wDRDF&wDRDF
	VIII	wF	wF	-	-
cBCC2	I	$< 0;0/1/- >_{a,v}$	$< 1;1/0/- >_{a,v}$	1PF2	CFst
		$< 1;1/0/- >_{a,v}$	$< 0;0/1/- >_{a,v}$	1PF2	CFst
	II	$< 1;1r1/0/0 >_{a,v}$	$< 0;0r0/1/1 >_{a,v}$	1PF2	CFrd
		$< 0;0r0/1/1 >_{a,v}$	$< 1;1r1/0/0 >_{a,v}$	1PF2	CFrd
	III	$< 1;1r1/0/1 >_{a,v}$	$< 0;0r0/1/0 >_{a,v}$	1PF2	CFdrd
		$< 0;0r0/1/0 >_{a,v}$	$< 1;1r1/0/1 >_{a,v}$	1PF2	CFdrd
	IV	$< 1;1r1 : 1r1/0/0 >_{a,v}$	$< 0;0r0 : 0r0/1/1 >_{a,v}$	2PF2	wCFrd&wRDF
		$< 0;0r0 : 0r0/1/1 >_{a,v}$	$< 1;1r1 : 1r1/0/0 >_{a,v}$	2PF2	wCFrd&wRDF
	V	$< 1;1r1 : 1r1/0/1 >_{a,v}$	$< 0;0r0 : 0r0/1/0 >_{a,v}$	2PF2	wCFdrd&wDRDF
		$< 0;0r0 : 0r0/1/1 >_{a,v}$	$< 1;1r1 : 1r1/0/0 >_{a,v}$	2PF2	wCFdrd&wDRDF
	V	$< 1;1r1 : 1r1 : 1r1/0/0 >_{a,v}$	$< 0;0r0 : 0r0 : 0r0/1/1 >_{a,v}$	3PF2	wCFrd&wRDF&wRDF
		$< 0;0r0 : 0r0 : 0r0/1/0 >_{a,v}$	$< 1;1r1 : 1r1 : 1r1/0/1 >_{a,v}$	3PF2	wCFrd&wRDF&wRDF
	VI	$< 1;1r1 : 1r1 : 1r1/0/0 >_{a,v}$	$< 0;0r0 : 0r0 : 0r0/1/0 >_{a,v}$	3PF2	wCFdrd&wDRDF&wDRDF
		$< 0;0r0 : 0r0 : 0r0/1/0 >_{a,v}$	$< 1;1r1 : 1r1 : 1r1/0/1 >_{a,v}$	3PF2	wCFdrd&wDRDF&wDRDF
	VII	wF	wF	-	-
cBCC3	I	$< 1/0/- >_v$	$< 0/1/- >_v$	1PF1	SF
	II	$< 1r1/0/0 >_v$	$< 0r0/1/1 >_v$	1PF1	RDF
	III	$< 1r1/0/1 >_v$	$< 0r0/1/0 >_v$	1PF1	DRDF
	IV	$< 1r1 : 1r1/0/0 >_v$	$< 0r0 : 0r0/1/1 >_v$	2PF1	wRDF&wRDF
	V	$< 1r1 : 1r1/0/1 >_v$	$< 0r0 : 0r0/1/0 >_v$	2PF1	wDRDF&wDRDF
	VI	$< 1r1 : 1r1 : 1r1/0/0 >_v$	$< 0r0 : 0r0 : 0r0/1/1 >_v$	3PF1	wRDF&wRDF&wRDF
	VII	$< 1r1 : 1r1 : 1r1/0/1 >_v$	$< 0r0 : 0r0 : 0r0/1/0 >_v$	3PF1	wDRDF&wDRDF&wDRDF
	VIII	wF	wF	-	-
dBCC1	I	$< 0;1/0/- >_{a,v}$	$< 1;0/1/- >_{a,v}$	1PF2	CFst
	II	$< 0;1r1/0/0 >_{a,v}$	$< 1;0r0/1/1 >_{a,v}$	1PF2	CFrd
	III	$< 0;1r1/0/1 >_{a,v}$	$< 1;0r0/1/0 >_{a,v}$	1PF2	CFdr
	IV	$< 0;1r1 : 1r1/0/0 >_{a,v}$	$< 1;0r0 : 0r0/1/1 >_{a,v}$	2PF2	wCFrd&wRDF
	V	$< 0;1r1 : 1r1/0/1 >_{a,v}$	$< 1;0r0 : 0r0/1/0 >_{a,v}$	2PF2	wCFdrd&wDRDF
	VI	$< 0;1r1 : 1r1 : 1r1/0/0 >_{a,v}$	$< 1;0r0 : 0r0 : 0r0/1/1 >_{a,v}$	3PF2	wCFrd&wRDF&wRDF
	VII	$< 0;1r1 : 1r1 : 1r1/0/1 >_{a,v}$	$< 1;0r0 : 0r0 : 0r0/1/0 >_{a,v}$	3PF2	wCFdrd&wDRDF&wDRDF
	VIII	wF	wF	-	-
dBCC2	I	$< 0;0/1/- >_{a,v}$	$< 1;1/0/- >_{a,v}$	1PF2	CFst
		$< 1;1/0/- >_{a,v}$	$< 0;0/1/- >_{a,v}$	1PF2	CFst
	II	$< 1;1r1/0/0 >_{a,v}$	$< 0;0r0/1/1 >_{a,v}$	1PF2	CFrd
		$< 0;0r0/1/1 >_{a,v}$	$< 1;1r1/0/0 >_{a,v}$	1PF2	CFrd
	III	$< 1;1r1/0/1 >_{a,v}$	$< 0;0r0/1/0 >_{a,v}$	1PF2	CFdrd
		$< 0;0r0/1/0 >_{a,v}$	$< 1;1r1/0/1 >_{a,v}$	1PF2	CFdrd
	IV	$< 1;1r1 : 1r1/0/0 >_{a,v}$	$< 0;0r0 : 0r0/1/1 >_{a,v}$	2PF2	wCFrd&wRDF
		$< 0;0r0 : 0r0/1/1 >_{a,v}$	$< 1;1r1 : 1r1/0/0 >_{a,v}$	2PF2	wCFrd&wRDF
	V	$< 1;1r1 : 1r1/0/1 >_{a,v}$	$< 0;0r0 : 0r0/1/0 >_{a,v}$	2PF2	wCFdrd&wDRDF
		$< 0;0r0 : 0r0/1/1 >_{a,v}$	$< 1;1r1 : 1r1/0/0 >_{a,v}$	2PF2	wCFdrd&wDRDF
	VI	$< 1;1r1 : 1r1 : 1r1/0/0 >_{a,v}$	$< 0;0r0 : 0r0 : 0r0/1/1 >_{a,v}$	3PF2	wCFrd&wRDF&wRDF
		$< 0;0r0 : 0r0 : 0r0/1/0 >_{a,v}$	$< 1;1r1 : 1r1 : 1r1/0/1 >_{a,v}$	3PF2	wCFrd&wRDF&wRDF
	VII	$< 1;1r1 : 1r1 : 1r1/0/1 >_{a,v}$	$< 0;0r0 : 0r0 : 0r0/1/0 >_{a,v}$	3PF2	wCFdrd&wDRDF&wDRDF
		$< 0;0r0 : 0r0 : 0r0/1/0 >_{a,v}$	$< 1;1r1 : 1r1 : 1r1/0/1 >_{a,v}$	3PF2	wCFdrd&wDRDF&wDRDF
	VIII	wF	wF	-	-

Index

address decoders 25-27
address decoder faults 11
address order 106, 107
add-hoc test 12

BISR 185, 193, 194, 196
BIST 11, 13, 185, 192, 196
bit lines 22
Butterfly 109
burn-in 5

CF (Coupling Fault) 11, 41, 13, 50-52, 95,96
CFdrd (Deceptive Read Destructive CF) 51, 96
CFds (Disturb CF) 50, 96
CFid (Idempotent CF) 13, 50, 108
CFin (Inversion CF) 51, 108
CFir (Incorrect Read CF) 52, 96
CFrd (Read Destructive CF) 51, 96
CFrr (Random Read CF) 52, 96
CFrrd (Random Read Destructive CF) 51
CFst (State CF) 11, 50, 95
CFtr (Transition CF) 51, 96
CFud (Undefined Disturb CF) 50
CFur (Undefined Read CF) 52
CFus (Undefined State CF) 50
CFwd (Write Destructive CF) 51
CFuw (Undefined Write CF) 52
characterization 5

DFT 12, 14, 15

DRAMs (Dynamic Random Access Memories) 8, 9
DRDF (Deceptive Read Destructive Fault) 46, 94
DRF (Data Retention Fault) 11, 47, 95

fault
    cell 132
    coupling 11, 41
    detection 4
    diagnosis 4
    dynamic 11, 39, 40, 189
    linked 39, 40
    MP (multi-port) 39, 41, 69
    multi-cell 39, 41
    notation 39, 42, 43, 47, 49, 53, 63, 93, 151
    port 132
    pPFs (p-port) 41, 159, 160, 162
    three-port 41, 151, 154, 156
    simple 39, 40
    single-cell 39, 41, 42, 49, 53, 55, 99, 109, 11, 113, 154, 155
    SP (single-port) 39, 41, 42, 49, 69, 93
    static 39, 40, 188
    two-cell 41, 42,49, 93, 58-62, 93, 98, 99, 110, 111, 113, 154, 155, 160, 162
    two-port 41, 53, 69, 97-99
    weak 13, 55, 56